Objective Agricultural Entomology

Objective Agricultural Entomology

Question Bank for for Ready Reference to the Students, Teachers and Researchers for Civil Service Examinations (State and Central), NET, SET, Ph.D. and Allied Examinations

— *Authors* —

A.S. Bagde

H.H. Powar

V.V. Pashte

2014

Daya Publishing House®

A Division of

Astral International Pvt. Ltd.

New Delhi – 110 002

ISBN 978-93-5130-226-1 (International Edition)

Publisher's note:
Every possible effort has been made to ensure that the information contained in this book is accurate at the time of going to press, and the publisher and author cannot accept responsibility for any errors or omissions, however caused. No responsibility for loss or damage occasioned to any person acting, or refraining from action, as a result of the material in this publication can be accepted by the editor, the publisher or the author. The Publisher is not associated with any product or vendor mentioned in the book. The contents of this work are intended to further general scientific research, understanding and discussion only. Readers should consult with a specialist where appropriate.
Every effort has been made to trace the owners of copyright material used in this book, if any. The author and the publisher will be grateful for any omission brought to their notice for acknowledgement in the future editions of the book.

Published by : **Daya Publishing House®**
A Division of
Astral International Pvt. Ltd.
– ISO 9001:2008 Certified Company –
81, Darya Ganj, Near Hindi Park,
Delhi Medical Association Road,
New Delhi - 110 002
Phone: 011-4354 9197, 2327 8134
Fax: +91-11-2324 3060
E-mail: info@astralint.com
Website: www.astralint.com

Laser Typesetting : **Classic Computer Services**

Preface

Entomology is one of the branch of agriculture, which having importance day by day. The general understanding of this subject must in simple way in new era. For various competitive examination of agriculture as well as civil examinations this subject to be simplified for better understanding of the student. In the market most of the books available in the form of multiple choice questions, in this book we are tried to simplify the questions in single answer which is simple to remember and easy to study.

This book included different topics of entomology such as morphology, taxonomy, toxicology, crop pest etc. this book will be great use to specially students appearing for JRF, SRF, NET/ARS as well as Ms.CET, State entrance exam. We acknowledge thanks my teachers, students, friends. Special thanks to Dr. A.G. Chandele, Prof. P.K. Dharne, Dr. N.G. Patil, Dr. P.B. Mohite, Dr. S.B. Kharbade, Agril. Asst. R.K. Patil, Siddhartha Kamble, Sagar Aher, Ganesh Misal, Mahesh Gaikwad, Chetan Pawar, Tanaji Patil who inspired to write this book. We also grateful to our family members for supporting us. It is my great pleasure to express my sincere thanks to Prof. A.B. Deshmukh, Librarian College of Agriculture, Pune and Astral International (P) Limited, New Delhi Publisher. Criticism and suggestions for the improvement of the book will be welcome.

A.S. Bagde

H.H. Powar

V.V. Pashte

Contents

Chapter 1
Syllabus for ARS-NET/SRF/JRF/IARI-Ph.D

SYLLABUS FOR ARS/NET EXAMINATION

Unit 1: Systematics

History and development of Entomology, Evolution of insects, position of insects in the animal world, characteristics of phylum Arthropoda, structural features of important arthropod groups such as Trilobita, Chelicerata and Mandibulata, structural features of important classes of phylum Arthropoda *viz.* Arachnida, Crustacea, Chilopoda, Diplopoda and Hexapoda. Classification of insects up to order level, habits, habitats and distinguishing features of different Order and important Families.

Unit 2: Morphology

Body wall, its structure, outgrowths, endoskeleton, Body regions, segmentation, sclerites and sutures. Head and head appendages, types of mouth parts, antennae, their structure and types. Thorax structure, thoracic appendages and their modification. Wings, their modification and venation, Abdomen; structure, abdominal appendages both in Pterygota and Apterygota. External genitalia, general structure and modification in important insect orders.

Unit 3: Embryology, Internal Anatomy and Physiology

Embryonic and post embryonic development, types of metamorphosis, physiology of ecdysis. General features and types of larvae and pupae. Structure, function and physiology of Digestive, Circulatory, Respiratory, Reproductive, Nervous and Excretory systems, Sense Organs; structure and types. Insect food and nutrition; minerals, carbohydrates, proteins and amino acids, lipids, vitamins and their role in growth and development, artificial diets.

Unit 4: Ecology

Concept of ecology, Environment and its components-biotic and abiotic factors and their effects on growth, development, population dynamics, distribution and dispersal. Principle of biogeography and insects biodiversity. Biotic potential and environmental resistance. Ecosystems, agroecosystems analysis, their characteristics and functioning. Intra and inter specific relationship; competition, predator-prey and host-parasite interactions, ecological niche. Life table studies, population models. Food chain and food web. Arthropod population monitoring, pest forecasting. Diapause and causes of pest out breaks.

Unit 5: Biological Control

Importance and scope of biological control, history of biological control: Biocontrol agents-parasites, predators and insect pathogens. Important entomophagous insect Orders and Families. Ecological, biological, taxonomic, legal and economic aspects of biological control, phenomena of multiple parasitism, hyperparasitism, superparasitism and their applied importance. Principles and procedures of using exotic biocontrol agents. Utilization of natural biocontrol agents: conservation, habitat management and augmentation. Mass multiplication techniques and economics. Effective evaluation techniques, Biocontrol organizations in world and India. Successful cases of biological control of pests. Entomophilic pathogens: bacteria, fungi, viruses, rickettsiae, Protozoan and nematodes, Modes of transmission, methods of uses, symptoms of infection. Microbial insecticides and their formulation. Merits and demerits of microbial control. Role of biocontrol agents and microbial insecticides in Integrated Pest Management.

Unit 6: Chemical Control and Toxicology

History, scope and principles of chemical control. Insecticides and their classification. Formulations of insecticides. Susceptibility of insects to the entry of insecticides. Physical, chemical and toxicological properties of different groups of insecticides: chlorinated hydrocarbons, organophosphates, carbamates, synthetic pyrethroids, chlordimeform, chitin synthesis inhibitors, avermectins, nitroguandines, phenylpyrrozzoles, botanicals (natural pyrethroids, rotenone, neem products, nicotine, pongamia spp. etc). Combination insecticides. Problems of pesticide hazards and environmental pollution. Safe use of pesticides, precautions and first aid treatments. Insecticides Act 1968, registration and quality control of insecticides. Evaluation of toxicity, methods of toxicity testing, determination of LD 50, LT 50, RL 50 etc. Pesticides residues in the environment and their dynamics of movements, methods of residue. Pharmacology of insect poisons. Mode of action of different groups of insecticides; neuroactive (axonal and synaptic) poisons, respiratory poisons, chitin synthesis inhibitors. Metabolism of insecticides; activative and degradative metabolism, detoxification enzymes and their role in metabolism. Selectivity of insecticidal actions; insecticide resistance; mechanism, genetics and management of insecticide resistance.

Unit 7: Host Plant Resistance

Chemical ecology: mechano and chemo receptors. Host plant selection by phytophagous insects. Secondary plant substances and their defenses against phytophagous insect. Basis of resistance (Antixenosis, Antobiosis, Tolerance). Biotypes development and its remedial measures. Tritrophic interactions, induced resistance. Breeding for insect resistant plant varieties. Resistance development and evaluation techniques. Genetics of Resistance: vertical resistance, horizontal resistance, oligogenic resistance, polygenic resistance. Biotechnological approaches and development of transgenic insect resistant plants, its advantages and limitations. Case histories. Insect resistance to transgenic plants and its management.

Unit 8: Innovative Approaches in Pest Control

Behavioral control: pheromones-types and uses, advantages and limitations. Hormonal control: types and function of insect hormones, insect hormone mimics, advantages and limitations. chemosterilants, antifeedants, attractants, repellents; their types, method of applications, advantages and limitations. Genetic control: concepts and methods, case histories, advantages and limitations. Potentialities of IPM.

Unit 9: Integrated Pest Management

History, concept and principles of IPM. Components of IPM: Host plant resistance, agronomic manipulations, mechanical and physical methods, chemical methods, biocontrol agents utilization, genetic and behavioral control strategy etc. IPM strategies for field and horticultural crops. IPM case histories. Concept of damage levels- Economic threshold levels (ETL), Economic injury levels (EIL) and their determination. System approach, Agro ecosystem and cropping system *vs.* IPM. Constraints and Strategies of IPM implementation.

Unit 10: Pesticide Application Equipments

Types of appliances: sprayers, dusters, fog generators, smoke generators, soil injecting guns, seed treating drums, flame throwers, etc. Power operated sprayers and dusters. Types of nozzles and their uses. Maintenance of appliances. Aerial application of pesticides, principles of aerial application, factors affecting the effectiveness of aerial application. Equipments for aerial applications. Advantages and disadvantages of aerial application.

Unit 11: Pests of Field Crops and their Management

Distribution, host range, biology and bionomics, nature of damage and management of arthropod pests of cereals, Oilseed, pulses and fibre crops, sugarcane and tobacco. Polyphagous pests: locusts, termites, hairy catepillars, cut worms and white grubs.

Unit 12: Pests of Horticultural Crops and their Management

Distribution, host range, biology and bionomics, nature of damage and management of arthropod pests of vegetables, fruits and plantation crops, spices, condiments and ornamentals.

Unit 13: Pests of Stored Products and their Management

Fundamentals of storage of grains and grain products. Storage losses, sources of infestation/infection, factors influencing losses, insect and non-insect pests, their nature of damage and control. Microflora in storage environment and their control. Storage structures, bulk storage and bag storage, their relative efficacy and demerits. Grain drying methods and aeration. Non-insect pests (rodents, birds, mites) of stored products and their control. Integrated management of storage pests.

Unit 14: Arthrpod Vectors of Plant Diseases

Common arthropod vectors *viz.*, aphids, leaf hoppers, plant hoppers, whiteflies, thrips, psylids, beetles, weevils, flies, bees and mites and their relationship with the plant pathogenic fungi, bacteria, viruses, mycoplasma. Mechanism of pathogen transmission : Active mechanical transmission, biological transmission. Toxicogenic insects, mites and phytotoxemia. Some important arthropod vector transmitted diseases and their epidemiology in India. Management of vector and its effect on control of diseases.

Unit 15: Honey Bees and Bee-keeping

Honey bees and their economic importance. Bee species, their behaviour, habit and habitats. Bee Keeping: bee pasturage, hives and equipments, seasonal management. Bee enemies including diseases and their control.

Unit 16: Silkworms and Sericulture

Silkworm species, their systematic position and salient features. Rearing techniques of mulberry-muga-eri and tassar silkworms. Nutritional requirements of silkworms. Sericulture: rearing house and appliances, silkworm breeds, principles of voltism and nioultism, seed production and its economics. Enemies and diseases of silkworms and their management. Sericulture organization in India.

Unit 17: Lac Insect

Lac insect, its biology, habit and habitats. Host Trees: pruning, inoculation, lac cropping techniques, and harvesting. Enemies of lac insect and their control.

Unit 18: Other Useful Insects

Pollinators, biocontrol agents of weeds, soil fertility improving agents, scavengers. Use of insects and insect products in medicines. Usefulness of insects in scientific investigations, insects as food.

Unit 19: Statistics and Computer Application

Frequency distribution, mean, mode and median. Standard, normal, bionomial and Poisson's distribution, Sampling methods and standard errors. Correlation and regression: Partial and multiple, tests of significance; t, F, chi- square, Duncan's multiple range tests. Design of experiments: Principles of Randomized block design, Completely randomized block design, Latin square design, Split-plot designs. Probit analysis. Use of soft ware packages like SPSS, SAS, etc. for the above tests and designs of experiments for analysis.

SYLLABUS FOR ICAR-JRF EXAMINATION

ENTOMOLOGY AND NEMATOLOGY

(Subjects: Agricultural Entomology, Nematology, Apiculture, Sericulture, Plant Protection.)

UNIT-I

Importance of Agriculture in national economy; basic principles of crop production; cultivation of rice, wheat, chickpea, pigeon-pea, sugarcane, groundnut, tomato, cole crops, mango, grapes, banana, oilseeds other than groundnut, soybean and mustard. Major soils of India, role of NPK and their deficiency symptoms. Mendelian genetics; elementary knowledge of photosynthesis; respiration, and transpiration; Major cropping systems (rice-wheat cropping, crop rotations, mixed cropping); soil degradation-soil salinity and acidity and management; some aspects of post-harvest technology; varietal improvement; importance of heterosis in crop production; crop protection principles in field and storage. Major insect pests and diseases of agricultural crops like rice, cotton, pulses, oilseed crops like groundnut, soybean and mustard, vegetables like tomato, cole crops; fruit crops like mango and banana and their management principles. Transgenic crops. Important rural development programmes in India; organizational set up of agricultural research, education and extension in India; Elements of statistics.

UNIT-II

Classification of animal kingdom up to class; distinguishing characters up to orders in class Insecta; general organization of an insect external morphology with special reference to lepidopteran larvae, coleopteran adults; and honeybee; metamorphosis and moulting; different physiological systems; insectplant relationship; insect pests of agricultural and horticultural crops, and their stored/ processed products, insect vectors of plant diseases- identification, biology, nature of damage, and their management tactics; and pests of household, medical and veterinary importance and their control; useful and beneficial insects like honeybee, lac insect, silkworm and pollinators; Nematode taxonomy, biology of important plant parasitic nematodes and their control; entomopathogenic nematodes, basic principles of insect and nematode pest management-cultural, biological, insecticidal, quarantine, and regulatory aspects; insecticide classification and insecticide resistance management; and insect protective transgenic crops.

SYLLABUS FOR ICAR SRF (PGS) EXAMINATION

PAPER-I

Section 1

- ☆ History and Geography of India—Events of significance with specific reference to scientific, cultural and economic life in India.
- ☆ General Awareness - Current events and analysis of their significance, books, authors, persons, awards, places, projects, sports and athletics.

- ☆ Famous personalities and their achievements in Arts/Science/Social life and trends of thought.
- ☆ Major development programmes in India relating to poverty alleviation, agriculture, rural development, etc. Five-Year Plans and their achievements—changes in the economic policies in the country and their impact on the agricultural scenario.

Section 2

- ☆ Agriculture in India - Basic information and data on the natural resources of the country; Agro-climatic and agro-ecological zones, Crops and their production; agricultural inputs and their use.
- ☆ Basic information on animal sciences and fisheries; Major breeds of livestock and their geographic distribution; Information on production and major recent achievements.
- ☆ Nutrition and agriculture - Basic information on foods and human nutrition.
- ☆ Quantitative analysis of data relating to agriculture/animal sciences/ fisheries. Drawing of conclusions from data, based upon reasoning and logic.
- ☆ Policies in India relating to science, industry, agricultural, forestry. World trade agreement and its possible impact on Indian agriculture.
- ☆ Current trends and achievements in science and scientific research on the global scene. Recent scientific achievements of significance in the country in major areas and their impact on agriculture, animal and fisheries.

PAPER-II
PLANT PROTECTION AND RELATED SCIENCES
[Agricultural Entomology, Nematology (Agriculture), Plant Pathology]
(AGRICULTURAL ENTOMOLOGY)

Section 1: Systematics

History and development of entomology. Position of insects in the phylum Arthropoda, relationship with other arthropods. Classification of insects up to order level. Habit, habitat and distinguishing characters of different orders. Distinguishing characters of families of economic importance. International code of zoological nomenclature. Body wall: structure, cuticular outgrowths, colouration and special integumentary structures. Body regions: sclerites and segmentation. Head: structure, types of mouth parts and antennae. Thorax sclerites and pterothorax. Wings, structure, venation, wing coupling and mechanism of flight. Legs: segmentation and types. Abdomen: segmentation and appendages. Genitalia and their modifications. Embryonic and post-embryonic development. Types of metamorphosis.

Section 2: Internal Anatomy and Physiology of Insects

Structure, modification and physiology of different systems, digestive, circulatory, respiratory, excretory, nervous, reproductive, musculature and insect sense organs (mechano-, hoto-and chemo-receptors). Physiology of moulting. Role of enzymes in growth and metamorphosis. Nutrition of insects, role of vitamins, proteins, carbohydrates, lipids and minerals. Extra and intra-cellular microorganisms.

Section 3: Ecology

Principles of ecology. Environment and its components. Effect of biotic and abiotic factors on bionomics, distribution, abundance and natural balance. Inter-and intra-specific relationships. Dispersal and migration. Life table studies. Population models. Surveillance and forecasting. Causes of insect outbreaks.

Section 4: Biological Control

Principles and scope of biological control. Use of parasitoids, predators and pathogens. Introduction of exotic natural enemies. Conservation and augmentation. Mass production techniques and economics. Field release and evaluation. Hostparasitoid relationships. Role of bacteria, fungi, protozoans and viruses; mode of action, mass production, formulation and field application.

Section 5: Insect-Host Plant Relationships

Chemical ecology. Tritrophic relation. Role of mechano- and chemoreceptors. Volatiles and secondary plant substances. Development of biotypes and measures to combat them. Breeding for insect-resistance in crop plants. Exploitation of wild species. Gene transfer. Case histories.

Section 6: Toxicology

Principles of toxicology. Classification of pesticides based on chemical structure, mode of entry, and mode of action. Toxicity and structure activity relationships. Mode of action and therapeutic methods for control of poisoning of arsenicals, fluoro compounds, dinitrophenol, plant products (rotenoids, nicotinoids, neem), chlorinated hydrocarbons, organophosphates, carbamates, natural and synthetic pyrethroids, cartap, avermectin, chloridimeform etc. Systemic insecticides. Phytoxicity. Compatibility, antagonism and synergism. Factors affecting toxicity of insecticides. Insecticide resistance and resurgence. Insecticides Act 1968. Registration and quality control of insecticides. Bioassay methods. Plant protection appliances. Pesticide formulations. Hazards of pesticides to humans, domestic animals and wild life.

Section 7: Insect Vectors of Plant Diseases

Role of insect vectors: aphids, leaf-hoppers, plant-hoppers, whiteflies, thrips, grass-hoppers and mites. Bacterial, viral, fungal and mycoplasma diseases and their transmission. Toxicogenic insects and phytotoxemia. Management of vectors.

Section 8: Pests of Crops and Stored Products and their Control

Biology, nature, extent of damage and control of pests of paddy, sorghum, maize, millets, wheat, barley, pulses, fruits, vegetables, cotton and other fibre crops, tobacco, oilseeds, sugarcane, ornamental plants, plantation crops such as tea, coffee, coconut, arecanut, cashew, cocoa, spices and condiments like pepper, cardamom, clove, nutmeg and chillies. Stored products pests. Detection and estimation of infestation. Management of stored product pests. Evaluation and efficacy of storage structures.

Polyphagous pests-locusts, termites, white grubs and hairy caterpillars and non-insect pests (mites, snails, birds and rats). Principles and components of integrated pest management. Economic thresholds. Economic injury levels. Action thresholds. Gain thresholds. IPM for important field and horticultural crops. Systems approach. Computer modelling. Case histories in IPM.

Section 9: Useful Insects

Honey bees, lac insect and silk worms.

Section 10: Statistics

Frequency distribution, means, median and mode, Standard, normal, binomial and Poisson's distribution, correlation, partial and multiple regression. Tests of significance: t, F and chi- quare test. Design of experiments. Basic principles of completely randomized, randomized block, latin square and split-plot designs, Probit analysis.

NEMATOLOGY (AGRICULTURE)

Section 1: History and Morphology of Nematodes

History and economic importance of nematology. Characters of phylum Nematoda, their taxonomic position and relationship with other organisms. General morphology. Anatomy- usculature, cuticle (finer) structures. Various systems—digestive, excretory, nervous system, etc. Reproduction and development of nematodes.

Section 2: Nematode Taxonomy

Principle and basis of taxonomy. Rules of nomenclature. Phylogeny. Nematode systematics. Classification of soil and plant-parasitic nematodes and their relationship. Detailed classification of plants-parasitics nematodes up to generic level with emphasis on economic importance.

Section 3: Nematological Techniques

Survey and sampling of nematodes. Methods of extraction of nematodes from soil and plant material. Microscopy, its principles and types. Methods of killing, fixing, preserving, staining and mounting of nematodes including TEM and SEM. Techniques for histopathology, histochemistry and culturing of nematodes including axenic methods. Experimental techniques for proving pathogenicity. Estimation of crop losses. Nematicide screening, residual analysis in plants and soil. Groundwater pollution. Screening of crop germplasm/genotypes for resistance.

Section 4: Plant Diseases Caused by Nematodes

Diseases caused by plant-parasitic nematodes-symptomatology, biology, distribution, economics and control of plant parasitic nematodes of economic importance (Pratylenchus, Radopholus, Hirschmanniella, Meloidogyne, Nacobbus, Heterodera, Globodera and other cyst forming nematodes, Rotylenchulus, Tylenchulus, Ditylenchus, Anguina, Aphelenchoides, Bursaphelenchus, Rhadinaphelenchus, Tylenchorhynchus, Xiphinema, Longidorus, Trichodorous, Paratrichodorus, Rotylenchus, Helicotylenchus, Hoplolaimus, Scutellonema, Criconematids etc.).

Section 5: Nematode Ecology

Ecological classification and dispersal of nematodes. Adaptations to parasitic mode of life. Soil as environment for nematodes. Effect of soil factors like temperature, moisture, porosity, soil type, organic matter, soil chemicals and pH on nematode survival, activity and multiplication. Effect of host on population of nematodes. Community analysis.

Section 6: Plant Nematode Relationships

Host-parasite relationships. Origin of parasitism among nematodes and types of parasitism. Injury caused by various groups of plant parasitic nematodes, mechanisms involved, and histopathology of nematode infected tissues. Physiological specialization and host specificity. Mechanisms of resistance in plants. Changes in physiology and biochemistry of plants due to nematodes.

Section 7: Nematode Physiology

Principles of nematode physiology, Nutrition, digestion and assimilation. Pseudocoelom and function of transport. Physiology of respiration, metabolism and excretion. Physiology of muscles, nervous and sensory responses and behaviour. Physiology of growth, reproduction and development. Role of pheromones in communication.

Section 8: Nematode Management

Principles and methods of nematode management-physical, cultural, biological, legislative and chemical methods. Nematicides—their formulations, applications and mode of action. Concept of host resistance for nematode control including their mechanisms. Integrated nematode management.

Section 9: Disease Complexes Caused by Nematodes and Other Pathogens

Interaction of nematodes with other nematodes (predatory and parasitic). Interaction with other disease causing agents like bacteria, fungi, viruses and mycorrhiza. Nematodes as biological indicators of toxic environmental contaminants.

Section 10: Statistics

Frequency distribution. Measures of central tendency and dispersion-mean, median, mode, standard deviation etc. Population distributions-normal, binomial and poisson. Correlations- artial and multiple. Regression—simple and multiple. Test of significance, t, F and Chi-square tests. Experimental designs—basic principles of randomized block, Latin square and split plot designs.

PLANT PATHOLOGY

Section 1: History and Principles of Plant Pathology

Milestones in phytopathology with special reference to mycology, bacteriology, virology and mycoplasmatology. History of development of epidemiological, physiological and molecular plant pathology. Development of chemical, legislative and biological protection measures including breeding for disease resistance. Nature and classification of plant diseases; Koch's postulates. Inter-relationships between various plant pathogens and their hosts, Growth, reproduction and dispersal of plant pathogens. Factors influencing infection. Entry of pathogens into plant. Colonization of infected plant, result of infection, and symptomatology.

Section 2: Laboratory and Analytical Techniques

Preparation and sterilization of common, media. Methods of isolation of plant pathogens. Purification using single spore, single hyphal tip isolation, and dilution/ streak planting. Purification of viruses. Maintenance and preservation of micro-organisms in pure culture, methods of inoculation. Laboratory equipments and their use: autoclave, hot air oven, laminar flow, pH meter, conductivity bridge, spectrophotometer, electrophoretic apparatus, light and electron microscopy Stains and staining. Selective media.

Section 3: Physiological and Molecular Plant Pathology

Altered metabolism of plants under biotic and abiotic stresses. Molecular mechanisms of pathogenesis, recognition phenomenon, penetration. invasion, primary disease determinant. Enzymes and toxins in relation to plant diseases. Mechanism of resistance, phytoalexins. Molecular detection of pathogens. High and low sugar pathogens. Tissue culture, somaclonal variation, somatic hybridization. Elementary genetic engineering.

Section 4: Mycology and Fungal Physiology

Economic mycology, edible fungi, and entomogenous fungi. Mycorrhizal associations. Fungal cell and its structure. Cell organelles, their morphology, functions and chemical composition, genetics and variation. Classification of fungi. Life cycles of fungi. Factors affecting spore formation, germination, metabolism, maturation and dormancy. Fungal nutrition and metabolism, role of macro/micro-elements and vitamins, antibiotics, fungal toxins and enzymes.

Section 5: Bacteriology and Virology

Morphology and structure of prokaryotic cell in relation to function. Origin, identification and classification of bacteria. Infrastructure and chemical composition; metachromatical detection, growth curve and nutrition. Bacterial genetics and variability, plasmid biology, transformation, conjugation, transduction, and transposon. Bacteriophages, ilytic and lysogenic cycles. Prokaryotic inhibitors and their mode of action. Economic uses of bacteria. Bdellovibrio, MLOs and other fastidious prokaryotes. Nature, composition and architecture of viruses and viroids. Properties of viruses. Inclusion bodies, and assay of plant viruses including biological, physical, chemical, serological and molecular methods. Techniques used in diagnosis. Behaviour of viruses in plants including infection, replication, movement and accumulation. Histopathological changes induced by viruses in plants. Transmission of viruses. Virus-vector relationships. Nomenclature and classification of viruses. Variability in viruses and viroids.

Section 6: Plant Disease Epidemiology

Concepts in epidemiology. Development of disease in plant population. Role of environment and meteorological factors in the development of plant disease epidemics.

Survey, surveillance and forecasting of diseases and assessment of losses. Epidemics analysis and prediction.

Section 7: Phanerogamic Diseases and Abiotic Stress

Diseases caused by phanerogamic parasites and their management. Abiotic stresses. Diseases due to unfavourable soil environment, atmospheric pollution and nutritional deficiencies.

Section 8: Fungal Diseases of Crops

Fungal diseases of cereals, millets, oilseeds and pulses, fruits and vegetables, and plantation, fibre, spices and ornamental crops with special reference to the etiology, disease cycle, perennation, epidemiology and management. Post-harvest diseases in transit and storage, and their management.

Section 9: Bacterial and Viral Diseases of Crop Plants

Crop diseases of cereals, pulses, oilseeds, vegetables, fruits, plantation and fibre crops caused by bacteria, fastidious prokaryotes, viruses and viroids. Mode of transmission, virus-vector relationships, epidemiology and management.

Section 10: Management of Plant-Diseases

Chemical, Biological and Legislative General principles of plant quarantine: domestic and international quarantine. Exotic pathogens and pathogens introduced into India. Disease resistance; genetic basis of diseases resistance, genetic basis of pathogenicity, gene and gene hypothesis, breeding for disease resistance, methods for screening for resistance in a segregating population. Tissue culture: production of disease-free plants, *in vitro* selection for disease resistance. Seed certification. Commercial production of disease-free seeds. Groups of fungicides and antibiotics. Mode of action of fungicides and antibiotics. Bioassay of fungicides. Compatibility of fungicides with other agricultural chemicals. Spraying and dusting of fungicides. Spraying and dusting equipments, their care and maintenance. Resistance and tolerance to fungicides. Influence of fungicides on environment. Integrated management. Micro-organisms antagonistic to plant pathogens in soil, rhizosphere and phyllosphere, plant growth-promoting bacteria (PGPB), deleterious rhizobacteria (DRB), hypovirulence, and cross protection. Use of microorganisms in the control of plant diseases.

Section 11: Statistics

Frequency distribution, mean, median and mode. Standard deviation. Normal, binomial and Poisson distributions. Probit analysis. Correlations, regression coefficients, and multiple regression. Tests of significance, t, F and Chi-square tests. Experimental designs basic principles, completely randomized, randomized block, Latin square and split-plot designs and their analysis.

SYLLABUS FOR IARI Ph.D. EXAMINATION

Position of insects in animal kingdom - their origin, phylogeny and distribution; history and basis of insect classification; distinguishing characters of insect Orders and economically important families; concept of species and speciation; rules and regulations of zoological nomenclature; morphology-external and internal; embryonic and post-embryonic development. Insect ecology-biotic potential, biotic and abiotic resistance, effect of temperature, humidity and light on insect development and population dynamics; diapause, food chain, migration and dispersal. Fundamentals of insect physiology, different systems, their structure and function, metabolism, sense organs, insect behaviour, host plant relationship. Social and other beneficial insects; pests of field crops and stored food; principles of pest control; classification, mode of action and metabolism of insecticides; insecticidal residues; resistance and resurgence; parasites, predators and pathogenic microoganisms of crop pests, biological control. Antifeedants, hormones, growth regulators, semiochemicals, host-plant resistance and genetic manipulation, insect quarantine; concept of integrated pest management; non-insect pests and their control

Chapter 2
Interesting Facts about Insects

☆ A particular Asian moth has evolved to feed on the tears of buffalo.

☆ Fireflies are beetles that use light to attract a mate.

☆ When a flea jumps it accelerates at a rate 20x faster than a rocket being launched into space.

☆ To avoid being eaten, stick insects hold their twig-lie bodies and legs at the correct angle on a branch and also they even sway like twigs when a breeze blows.

☆ Malaysian giant stick insect eggs are just 4mm wide, but when the baby stick insects hatch they are already 28 cm long.

☆ The fastest known insect is a dragon fly that has been clocked at 58 kilometres an hour.

☆ The heaviest insect in the world is the Goliath beetle from Africa. A big male can weigh up to 100 grams.

☆ A flea can jump 130 times its own height.

☆ Fleas jump 200 times their body length, and grasshoppers can jump 80 times of their length.

☆ Some mosquitoes can beat their wings 600 times a second. A midge fly holds the record, with a wing beat of 1046 times a second.

☆ Butterflies get their name from the yellow brimstone butterfly of Europe that is first seen in the early spring or "butter" season.

☆ The atlas moth, one of the largest silk moths, can be mistaken for a medium-sized bat when flying.

- ☆ Butterflies and moths are found on all land masses except Antarctica.
- ☆ Praying mantids are among the few insects which can rotate their heads so they can literally look over their shoulders, making them extremely effective predators.
- ☆ The metallic-coloured wing covers of some beetles are used for jewellery.
- ☆ Female Queen Alexandra butterflies, from Papua and New Guinea, are the largest in the world, some with wingspans larger than 26 cm.
- ☆ The Jungle Nymph Stick is one of the heaviest insects. In Malaysia they are often kept by people who feed them guava leaves and use the droppings to make tea.
- ☆ A dragonfly has a lifespan of 24 hours.
- ☆ The common Black Ants and Wood Ants have no sting, but they can squirt a spray of formic acid. Some birds put ants in their feathers because the ants squirt formic acid which gets rid of the parasites.
- ☆ A species of earthworm, "*Megascolides australis*", in Australia can grow up to fifteen feet in length.
- ☆ The Honey Badger can withstand hundreds of African bee stings that would kill any other animal.
- ☆ Monarch caterpillars shed their skin four times before they become a chrysalis, growing over 2700 times their original size.
- ☆ The honeybee has to travel an average of 43,000 miles to collect enough nectar to make a pound of honey.
- ☆ Only male crickets can chirp.
- ☆ Pea crabs (the size of a pea) are the smallest crabs in the world.
- ☆ Ants carry the caterpillars of the large blue butterfly into their nest. There, the ants feed on the sweet honeydew produced by the caterpillars, which in return feed on the tiny ant larvae.
- ☆ The term "Honeymoon" comes from the middle ages, when a newly married couple was provided with enough honey wine to last for the first month of their married life.
- ☆ To survive the cold of winter months, many insects replace their body water with a chemical called glycerol, which acts as an "antifreeze" against the temperatures.
- ☆ The queen of a termite colony may lay 6000 to 7000 eggs per day, and can live for 15 to 50 years.

- ☆ Cicadas may live as long as 17 years and male cicada makes the loudest sound in the insect world (107 decibels) and the sound made by them can carry up to a mile.
- ☆ Ants can carry objects that weight 100 times the weight of their own body.
- ☆ Stick insect eggs that resemble hard seeds have a special, fatty capsule called a capitulum at one end. Ants enjoy the nutritional boost provided by the capitulum, and carry the stick insect eggs back to their nests for a meal. Once the ants feed on the fats and nutrients, they toss the eggs onto their garbage heap where they continue to incubate safe from predators. As the nymphs hatch, they make their way out of the ant nest.
- ☆ Giant Burrowing Cockroach (Australia): They are the largest cockroach in the world.
- ☆ Queen Alexandra's Birdwing: Papua New Guinea: It's the largest butterfly in the world. The first specimen ever found was taken down using a shotgun.
- ☆ Insect with shortest lifespan: Mayflies have the shortest lifespan of any insect. Although they spend two to three years as nymphs in the water, they live as little as one hour as adults, just long enough to mate and lay eggs.
- ☆ The oldest known insect on earth is the cockroach. Cockroaches date back 350 million years ago. They are extremely adaptable and can live everywhere from high in the mountains, to deep within the rain forest.
- ☆ By far, locusts are the most destructive insect. They consume their weight in food each day. When swarms hit crops, the crops are destroyed as there's nothing left. Swarms can contain up to a billion locusts.
- ☆ The smallest insect in the world is the Fairyfly. Its amazing size is a mere 0.01 inch. There are 1,424 known species of Fairyfly. It belongs to family Mymaridae. This creature is also commonly called Fairy Wasp. It is also the smallest known flying insect.
- ☆ The nine-spotted ladybug (*Coccinella novemnotata*) was designated the official state insect symbol of New York in 1989.

Chapter 3
History of Entomology

Multiple Choice Questions

1. ___________was done the first detailed classification of insects.
   (a) Umaswati (b) Charaka
   (c) Sushruta (d) Linnaeus

2. ___________done the classification of bees.
   (a) Umaswati (b) Charaka
   (c) Sushruta (d) Linnaeus

3. ___________wrote the book "Systema Naturae" in 1758 *i.e.* 10th Edition.
   (a) Carl Linnaeus (b) Sushruta
   (c) J. C. Fabricius (d) Darwin

4. ___________is the first entomologist who done study of Indian insects.
   (a) Carl Linnaeus (b) Dr. Horsfield
   (c) J. C. Fabricius (d) Umaswati

5. The ' Asiatic Society of Bengal' started at ________in _________year.
   (a) Calcutta, 1785 (b) Madras, 1785
   (c) Calcutta, 1885 (d) Madras, 1885

6. ___________is called first keeper of 'East India Museum'.
   (a) J. C. Fabricius (b) Carl Linnaeus
   (c) Dr. Horsfield (d) W. T. Blandford

7. **Bombay Natural History Society was started in __________year.**
   (a) 1886 (b) 1887
   (c) 1883 (d) 1783

8. **__________give the "Theory of Evaluation".**
   (a) Buchanan, 1859 (b) Darwin, 1859
   (c) Lionel de Nicevelle, 1901 (d) Maxwell Lefroy, 1859

9. **__________started the 'Fauna of British India' series as editorship in 1883.**
   (a) Lionel de Nicevelle (b) Maxwell Lefroy
   (c) W. T. Blandford (d) Buchanan

10. **__________is the first entomologist to the Government of India in 1901.**
    (a) Lionel de Nicevelle (b) Bartlett
    (c) Sushrutha (d) Donovan

11. **__________was called the first imperial entomologist.**
    (a) Maxwell lefroy (b) Geier
    (c) Dr. J. Aderson (d) Dr. Horsfield

12. **__________ has published the books "Indian Insect Pests" and "Indian Insect Life" in 1906 and 1909 respectively.**
    (a) Bartlett (b) W. T. Blandford
    (c) Carl Linnaeus (d) Maxwell Lefroy

13. **In the year __________destructive insect and pest act (DIPA) was enforced.**
    (a) 1914 (b) 1915
    (c) 1930 (d) 1931

14. **In__________year plant quarantine act was enforced.**
    (a) 1912 (b) 1913
    (c) 1914 (d) 1969

15. **Zoological Survey of India formed in the year__________.**
    (a) 1914 (b) 1916
    (c) 1912 (d) 1915

16. **Entomological Society of India formed in __________ year.**
    (a) 1914 (b) 1938
    (c) 1912 (d) 1969

17. **In the year__________ Locust Warning Organization was established.**
    (a) 1940 (b) 1939
    (c) 1965 (d) 1999

18. Directorate of Plant Protection Quarantine and Storage was established at ___________

(a) Faridabad (U.P.), 1945
(b) Faridabad (U.P.), 1946.
(c) Banglore (KA), 1946
(d) Banglore (KA), 1945

19. The word 'Entomology' derived from___________.

(a) Latin
(b) French
(c) Greek
(d) None of these

20. ___________ coined the term 'Integrated Pest Contol'.

(a) Bartlett
(b) Buchanan
(c) Ziedler
(d) Burrell

21. The term 'Pest Management' given by ___________ entomologist.

(a) Donovan
(b) Buchanan
(c) Bartlett
(d) Geier

22. Plant Protection and Quarantine Act was enforced in year___________.

(a) 1915
(b) 1914
(c) 1939
(d) 1940

23. ___________classified the Ants (pipilika) into six groups.

(a) Sushrutha
(b) Charaka
(c) Linnaeus
(d) Dr. Kerr

24. In 1782 ___________ published an account of Lac Insect.

(a) Dr. J. Anderson
(b) Dr. J. G. Koenig
(c) Dr. Kerr
(d) Linnaeus

25. ___________published the Natural History of Insects.

(a) Donovan
(b) Dr. J. Anderson
(c) Dr. Kerr
(d) T. B. Fleture

26. Central Insecticidal Act was enacted by Govt. of India in ___________ year and came into force from___________year.

(a) 1969 and 1st Feb., 1970
(b) 1968 and 1990
(c) 1968 and 1st Jan., 1971
(d) 1969 and 1972

27. The evolution of land plants in ___________ geological period.

(a) Jurassic
(b) Devonian
(c) Quaternary
(d) Carboniferous

28. In __________ geological period phytophagous insects are known.

(a) Carboniferous (b) Jurassic
(c) Devonian (d) Quaternary

29. The first record of winged insect was done in __________ period.

(a) Lower carboniferous (b) Upper carboniferous
(c) Upper jurassic (d) Quaternary

30. The first record of fossil insect __________ from Devonian period of Scotland.

(a) *Rhyniella praecursor* (b) *Rhodnius prollixus*
(c) *Dysdercus koenigii* (d) *Atherigona socata*

31. *Rhyniella praecursor* belong to order __________.

(a) Collembola (b) Coleoptera
(c) Lepidoptera (d) Heteroptera

32. The book "Silent Spring" was written by __________ in year 1962.

(a) Imms (b) Linnaeus
(c) Rachel Carson (d) Watson

33. Genetic engineering approval committee comes under the ministry of __________.

(a) City development (b) Finance
(c) Agriculture (d) Environment and Forests

34. Ecology term coined by __________.

(a) A. G. Tansley (b) Earnest Haekel
(c) Carl Linnaeus (d) None of these

Answer Key

1–a	7–c	13–a	19–c	25–a	31–a
2–b	8–b	14–a	20–a	26–c	32–c
3–a	9–c	15–b	21–d	27–b	33–d
4–c	10–a	16–b	22–b	28–a	34–b
5–a	11–a	17–b	23–a	29–b	
6–c	12–d	18–b	24–c	30–a	

Chapter 4
Taxonomy

1. Insect identification keys are primarily based on differences and similarities in insect morphology.
2. The level of taxonomy concerned with characterization and naming of a species is alpha taxonomy.
3. The level of taxonomy concerned with the arrangement of species in the biological hierarchy is beta taxonomy.
4. The level of taxonomy concerned with the studies of intra-specific populations to understand speciation, evolution rates and trends is gamma taxonomy.
5. Insects are grouped in to phylum: Arthropoda.
6. Insects are grouped in to sub phylum: Uniramia.
7. Insects belongs to class: Hexapoda.
8. The class of phylum Arthropod was an extinct: Trilobita.
9. Hetroptera and Homoptera are sub order of Hemiptera order.
10. Orthoptera order contents/have sub orders: Caelifera and Ensifera.
11. Caddis flies belongs to order: Tricoptera.
12. In which class of arthropods the body is divided into two regions *i.e.* cephalothorax and abdomen: Crustacea.
13. The class of arthropods in which the legs are unjointed: Onychophora.
14. Ant lions belong to the order Neuroptera.

15. The Chloropidae are a family of flies commonly known as frit flies or grass flies.
16. "Millipede" is a compound word formed from the Latin roots <u>*mille*</u> ("thousand") and <u>*pes*</u> ("foot").
17. The insect order Zoraptera contains a single family, the Zorotypidae.
18. The angel insect belongs to order Zoraptera.
19. The name Zoraptera, given by Filippo Silvestri in 1913.
20. Vein less bearing hymenopterans parasite belong to the family platygastridae.
21. The Thysanoptera were first described in 1744 as a genus *Physapus* by De Geer, and then renamed 'Thrips' by Linnaeus in 1758. In 1836 Haliday promoted the genus to the taxonomic rank of order, renaming them Thysanoptera.
22. Entognathus (mouth parts retracted within the head): Orders- Diplura, Protura and Collembola.
23. Ectognathus mouth parts (mouthparts are exposed): Order- Thysanura and Pterigota.
24. Proturans: sometimes referred to as "coneheads" and they do not have eyes or antennae.
25. In Terebrantian thrips, a single pupal instar follows, whereas in the Tubulifera, two pupal stages.
26. Alderflies and dobsonflies belong to order Megaloptera.
27. Snakeflies belongs to order Raphidioptera.
28. Order Raphidioptera: rhaphis = a needle; pteron = wing.
29. Beneficial insects largely belong to order hymenoptera.
30. Gladiators are called as rock crawlers, heelwalkers, mantophasmids and mantos.
31. Mantophasmatidae is suborder of carnivorus African insects discovered in 2001.
32. *Grylloblatta campodeiformis* the first species of Grylloblattidae (ice bugs or ice crowlers).
33. *Grylloblatta campodeiformis*, which means "cricket- cockroach shaped like a *Campodea*" (a kind of two-pronged bristletail).
34. The optimal living temperature for ice bug is between 1-4°C (33.8-39.2°F.

35. The wings of butterfly and birds are homologous organs.
36. The striking difference in male and female cockroach is: presence or absence of anal style.
37. Palaeopteran orders (Odonata and Ephimeroptera): No wing flexing mechanism.
38. An intermediate population unit which exists between species and deme is called a race.
39. The term species established by John Ray.
40. According to ICZN, super family ends with 'oidea'.
41. According to ICZN, family ends with 'idea'.
42. According to ICZN, sub family ends with 'inae'.
43. According to ICZN, tribe ends with 'ini'.
44. Largest order: Coleoptera.
45. Orders absent in India: Grylloblatodea and Zoraptera.
46. Smallest order: Zoraptera (ICAR-SRF, 2013).
47. Morphological species concept: separate species based on differences in appearance.
48. Biological species concept: separate species based on ability to interbreed.
49. Stone flies used as indicators of water purity.
50. Structurally more specialized order: Diptera.
51. Order containing single genus: Zoraptera.
52. Exopterygote insect order with pupal instar in the life cycle: Thysanoptera.
53. Cohort is present between: class and order.
54. Permothemis: connecting link between Ephemeroptera and odonata.
55. Lepismatidae (Thysanoptera): conneting link between Apterygota and Pterygota.
56. Termite texa without true workers: Mastotermitidae, Kalotermitidae and Termopsidae.
57. Termite texa without true soldiers: Anoplotermes and Apeculitermus.
58. The order Archaeognatha (sometimes called Microcoryphia) includes insects commonly known as jumping bristletails.

59. Froghopper belongs to family: Cercoptidae.
60. Males are winged and wingless females in: Embioptera, strepsiptera and coccids.
61. Order Archaeognatha: mandibles connect with the head capsule in only one place (monocondylic).
62. The mandibles move only in a transverse direction: order Thysanura.
63. In mayflies and damselflies, the gills are leaf-like in shape and located on the sides or rear of the abdomen.
64. Stoneflies and caddisflies have filamentous gills on the thorax or abdomen.
65. Uniramia includs onychophora, millipedes, centipedes, pauropods, symphylans and insects.
66. Myriapod includes both Chilopoda (centipedes) and Diplopoda (millipedes).
67. Robber flies belong to family: Asilidae.
68. Fly with beard face: Asilidae.
69. The Crabs have two pairs of antenna.
70. Beautiful order of insects: Lepidoptera.
71. Anal horns in the larval stage are found in family: Bombycidae.
72. Holotype: When a single specimen is clearly designated in the original description, this specimen is known as the *holotype* of that species. The holotype is typically placed in a major museum, or similar well-known public collection, so that it is freely available for later examination by other biologists.
73. Paratype: When the original description designated a holotype, there may still be additional specimens listed in the type series and those are termed paratypes.
74. Neotype: A neotype is a specimen later selected to serve as the single type specimen when an original holotype has been lost or destroyed or where the original author never cited a specimen.
75. Syntype: A syntype is any one of two or more specimens that is listed in a species description where no holotype was designated.
76. Lectotype: A lectotype is a specimen later selected to serve as the single type specimen for species originally described from a set of syntypes.

77. Paralectotype: A paralectotype is any additional specimen from among a set of syntypes, after a lectotype has been designated from among them.

78. Hapantotype: A special case in Protistans where the type consists of two or more specimens of "directly related individuals representing distinct stages in the life cycle"; these are collectively treated as a single entity, and lectotypes cannot be designated from among them.

79. Notonectidae is a cosmopolitan family of aquatic insects in the order Hemiptera, commonly called backswimmers because they swim upside down.

80. Paraphyletic: referring to a taxonomic group containing a single ancestor and some, but not all, of its descendants.

81. Analogous features: Morphological characteristics of two or more taxa that are superficially similar but not as a result of any common evolutionary origin. (*E.g.* birds, bats, and butterflies all have wings, but this is not because they are closely related).

82. Homologous features: Morphological characteristics of two or more taxa that indicate a common evolutionary origin, even though the organisms may differ in terms of other morphological features. (*e.g.* limb- common to many birds and most mammal), which indicates a common ancestor.

83. Distinctive 'eye-caps' formed by the basal segment of each antenna: Family nepticulidae and phyllocnistidae.

84. The mostly accepted insect classification: Imms, 1961.

85. The order in which both the pairs of wings are similar and membranous: Isoptera.

86. The broad classification of animal kingdom proposed by Hymen.

87. Rolling as protetive mechanism is seen in millipeds (ICAR-NET-2013).

88. Megaloptera and Planipennia are the suborders of Neuroptera.

89. Stingless bee belongs to Meliponini tribe.

90. Two antennal pair and five pairs of legs: crustaceans.

91. Butterfly: sex- determination- female heterogamy.

92. Butterfly female genome: ZW (ICAR-SRF, 2013).

93. Butterfly male genome: ZZ.

94. Psedohalteres: front wings of Strepsiptera.

95. Jhonston's organ: absent in Collembola and Diplura.

Fill in the Blank Type Questions with Answers

1. **Poisonous claws are present on legs of ________class of insects.**
   Ans: Chilopoda

2. **Antenna is absent in ________ class of arthropods.**
   Ans: Arachnida

3. **Non-mandibulated type of mouth parts is a characteristic of ________ and ________ classes of insects.**
   Ans: Onychophora and Arachnida

4. **Chelicerae are present in ________ class of arthropods.**
   Ans: Arachnida

5. **Cutaneous type of respiration done in ________ class of arthropods.**
   Ans: Onychophora

6. **Anamorphosis present in ________ and ________ class of phylum Arthropoda.**
   Ans: Onychophora and Crustacea

7. **Which of the following orders belongs to apterygota sub-class?**
   Ans:Thysanura, Collembolla, Diplura and Protura

8. **Which of the following orders have entognathus type of mouth parts?**
   Ans: Diplura, Collembolla and Protura

9. **Insects belong from Thysanura and Pterygota have________ mouthparts.**
   Ans: Ectognathus

10. **Pregenital abdominal appendages present in ________ insects.**
    Ans: Apterygota

11. **Scorpion respire by________.**
    Ans: Book lungs

12. **________ are paleoptern orders.**
    Ans: Ephemeroptea and Odonata

13. **________insect have body clothed with glistering scales.**
    Ans: Sliver fish

14. **________pest feeds on old books and there bindings.**
    Ans: Sliver fish

15. **In ________ order insect have six abdomenal segments and lack of compound eyes.**
    Ans: Collembolla

16. **Spring tail posses the callophore, tenaculum, furcula on which abdominal segments respectively?**

    Ans: 1st, 3rd and 4th segments

17. **Callophore or glue peg act as organ for ________.**

    Ans: Respiration, Adhesion and Water Absorption

18. **Pseudoculi is a minute structure on head of ________ insect.**

    Ans: Proturans

19. **Metamorphosis absent in ________order.**

    Ans: Thysanura, Collembola and Diplura

20. **Anamorphosis present in________where increase in number of abdominal segments at moulting.**

    Ans: Protura (Telsontail)

21. **________ is order of leaf and stick insects.**

    Ans: Phasmida

22. **In________, compound eyes and ocelli are absent.**

    Ans: Diplura

23. **Sub-imago and imago are the forms of adult found in ________ insect.**

    Ans: Mayfly (Ephemeroptera)

24. **Order Ephemeroptera have ________character.**

    Ans: Absent wing flexing mechanism, bilateral abdominal gills and Atrophied mouth parts

25. **Dragonflies and Damselflies belongs to sub order________ and ________ respectively.**

    Ans: Anisoptera and Zygoptera

26. **Mask (elongated Labium with hooks) and mula (form by fusion of Gulea and Lacinia) are structures found in order________.**

    Ans: Odonata

27. **________of compound eyes are present in Dragonflies.**

    Ans: Holoptic type

28. **Wing flexing mechanism absent in order________.**

    Ans: Odonata

29. **Male genitalia are asymmetrical in ________.**

    Ans: Grylloblattodea

30. **Pterostigma or stigma is a characteristic pigmented cell found on wings of _________.**
    Ans: Odonata

31. **Production of drumming sound characteristic of adult of _________.**
    Ans: Plecoptera.

32. **Pronymphs (first instar nymphs covered with loose cuticle) present in _________.**
    Ans: Orthoptera

33. **Locust is _________in nature.**
    Ans: Polymorphic

34. **In_________family males stridulate during night.**
    Ans: Gryllidae (Cricket)

35. **In mole cricket tymphanum is present on_________.**
    Ans: Tibia of forelegs

36. **_________insect is known as living tree thermometer.**
    Ans: Snow tree cricket

37. **In grasshopper tymphanum is present on _________abdominal segment.**
    Ans: First

38. **Regenerating type of legs are found in _________ order. (ICAR- SRF-2011)**
    Ans: Phasmida

39. **Wing venation mimics leaf venation in_________ insects.**
    Ans: Leaf insect

40. **A distinct 'Y' shaped epicranial suture is found in_________order.**
    Ans: Dermaptera (Earwing)

41. **Cerci are highly sclerotised and forcep like of_________insect.**
    Ans: Earwing

42. **_________insect have greatly enlarged basitarsus with silk gland.**
    Ans: Embiids

43. **In_________order the insect legs are modified to running backword.**
    Ans: Embioptera

44. **Cockraoch eggs are contained in _________.**
    Ans: Ootheca

45. **Fenestrae (*i.e.* degenerated ocelli) are present in _________.**
    Ans: Cockroach

46. **Eggs of mantids present in __________.**
Ans: Spumaline

47. **Cockroaches have pronotum like a __________.**
Ans: Shield

48. **Cockroach belongs to order __________.**
Ans: Dictyoptera

49. **Termites belong to order__________.**
Ans: Isoptera

50. **Dwelling place of termite is called__________**
Ans: Mound or Termitoria

51. **The trophallaxis (food sharing between members of species) observed in __________.**
Ans: Termites

52. **__________have symbiotic association with termite that helps in digestion of cellulose.**
Ans: Flagellated protozoa

53. **Nematocera, Brachycera and Cyclorhapha are sub orders of __________.**
Ans: Diptera

54. **Number of stylets present in the mouth parts of culicidae (Mosquito): __________**
Ans: Six

55. **__________ of mosquitoes is parasitic in nature.**
Ans: Females

56. **Terminal abdominal segments of housefly form__________ovipositor.**
Ans: Pseudo

57. **Viviparous type of reproduction observed in__________.**
Ans: Hippoboscidae (Dog fly)

58. **Forewings small club-like structure in male of __________.**
Ans: Strepsipterae (Stylopids)

59. **In order__________, mesothorax and metathorax are fused with first abdominal segment.**
Ans: Hymenoptera

60. **Forewings are not used in flight in__________ order.**
Ans: Coleoptera

61. **__________is largest order of insect.**
   Ans: Coleoptera

62. **Compuond eyes divided, and seen to be two pair of compound eyes in__________.**
   Ans: Gyrinidae (Whirlingig beetle)

63. **In bruchidae (Pulse beetle) the elytra do not cover the abdomen and some part is exposed that is called as__________.**
   Ans: Pigidium

64. **Serrated antenna found in __________.**
   Ans: Buprestidae (Jewel beetle)

65. **Mustard saw fly belongs to family__________ of sub order symphyta.**
   Ans: Tenthredinidae

66. **Apocrita, a sub order of Hymenoptera includes generally__________insects.**
   Ans: Parasitic

67. **__________type of pupa found in Lepidoptera.**
   Ans: Obtect

68. **Large numbers of Lepidopteran insects are grouped in sub order__________.**
   Ans: Ditrysia

69. **Osmeteria are present on the caterpillar of __________.**
   Ans: Papilionidae (Citrus Butterfly)

70. **Newly hatched nymph (naiads) is called __________in odonata.**
   Ans: Pronymph

71. **"Jet propulsion" is present in the naiads of __________.**
   Ans: Dragonfiles

72. **Berlese funnel is used to catch __________.**
   Ans: Soil dewling insect.

73. **The chemicals are used to kill insects__________.**
   Ans: Pottassium cyanide, Chloroform and Ethyl acetate

74. **Order __________is completely herbivorous.**
   Ans: Phasmida

75. **Order __________ of insect comprises largest insect species.**
   Ans: Coleoptera

Chapter 5
Morphology

1. Camel crickets have a characteristic hump-backed appearance.
2. German cockroach: *Blattella germanica*: two dark brown stripes on the body region (pronotal shield) just behind the head.
3. American cockroach: *Periplaneta Americana*: center portion of the pronotal shield is light brown, while the outer edges are yellow.
4. Brownbanded cockroach: *Supella longipalpa*: two lighter colored bands running across the body (located at the base of the wings and on the abdomen).
5. Vinegar flies have red eyes.
6. The green lacewing eggs are oval and secured to the plant by long slender stalk.
7. Stalk-eyed flies are insects of the fly family Diopsidae. (Possession of "eyestalks": projections from the sides of the head with the eyes at the end).
8. The mites have no antennae, compound eyes or wings.
9. The body of a mite is composed of just two main sections- a gnathosoma (which bears a pair of segmented pedipalps and the mouthpart) and a sac-like idiosoma (which bears the legs).
10. Females of the genus Phobaeticus are the world's longest insects.
11. Millipedes are arthropods that have two pairs of legs per segment (except for the first segment behind the head which does not have any appendages at all, and the next few which only have one pair of legs).
12. Uniquely among insects, mayflies possess paired genitalia, with the male having two penises and the female two gonopores.

13. The abdomen of mayflies is roughly cylindrical, with 10 segments and two long cerci at the tip.
14. Protura: The mouthparts are entognathous (enclosed within the head capsule) and consist of thin mandibles and maxillae.
15. Hind tibia with a large apical movable spur is characteristic feature of Delphacidae family.
16. Pronation: A classification of wings; *i.e.*, twisting, anterior margin downward, posterior margin upward.
17. Eight pairs of legs found in: mustard saw fly larvae.
18. Number of stylets in female mosquito: six.
19. Number of stylets in head louse: three.
20. Number of stylets in plant bug: four.
21. Number of stylets in Mallophaga: absent.
22. Accessory gland: A gland associated with reproductive organs of either males or females and producing substances accompanying the sperms or eggs.
23. Active space: The zone of pheromone concentration within which a response is elicited.
24. Aeropile: The opening in the chorion (egg shell) through which air enters, often covered by a plastron.
25. Allatectomy: An operation resulting in removal of the corpora allata.
26. Allometric growth: A genetically determined tendency for a certain body part to grow at a more rapid rate than other parts.
27. Aerostats: a pair of large air sacs at base of abdomen in Diptera.
28. Crook: the hook or recurved tip of the antenna in Hesperidae.
29. Basalare: A small sclerite in the upper part of the pleuron that articulates with the axillary sclerites.
30. The body of insect is monoblastic.
31. Spiracles are absent on prothorax.
32. All primitive insects have ectognathous mouthparts (mandibles and maxillae not enclosed within a head cavity).
33. Flagellomere: each segment of flagellum.
34. The main function of tentorium: to protect the brain.

35. In grasshopper one large plate is found at the end of 10th segment or tergites of abdomen.
36. Epistomal suture is absent in: *Blatta orientalis.*
37. Paired gonopore is present in: Ephemeroptera.
38. Asymetrical cerci is present in: Embioptera.
39. Saw like ovipositor in thysanoptera: sub order Terebrantia.
40. Exoloma: the apical margin of the wings.
41. Chemical formula of Chitin : $(C_8H_{13}O_5N)_n$.
42. Non cellular part of insect: Cuticle (ICAR-NET 2013).
43. Lepidopteran family which has two pairs of functional legs: nymphalidae.
44. Opisthogenital gonopore present in the terminal segment: Chillopoda.
45. Tympanal organs are present on the 1st abdominal segment: Acridoidea.
46. Pearmans organ (on coxa which acts as stridulatory organ) present in order: Psocoptera.
47. The males are winged and females are wingless: Coccids, Embioptera, Strepsiptera.
48. Complete apterous condition: Pscoptera, Mallophaga and Siphonaptera.
49. Cremaster: spine like structure found at the tip of abdomen.
50. Mesostigma: in Odonata, the spiracles of second thoracic segment.
51. Muller's thread: the common terminal thread of all the ovarian tubes.
52. Stomatotheca: that part of the pupa covering the mouth structures.
53. Stethidium: the entire thorax with all its appendages.
54. Axillary sclerites are absent in odonata.
55. Abdominal gills: respiratory organs found in the nymphs (naiads) of certain aquatic insects.
56. In Ephemeroptera (mayflies), paired gills are located along the sides of each abdominal segment.
57. Collophore: a fleshy, peg-like structure found in Collembola on the ventral side of the first abdominal segment.
58. Supraanal plate is also called as 11th tergite.
59. Number of spiracles are functional in grasshopper: 10.
60. Pteralia is part of the wing base. (IARI Ph. D. 2011).

61. Patagia: a pair of small vesicular organs situated at the bases of the anterior wings of lepidopterous insects. (IARI Ph. D. 2011).
62. The junction between thorax and abdomen is very narrow (petiolate): wasp-waist.
63. Each segment of the abdomen consists of a dorsal sclerite, the tergum, and a ventral sclerite, the sternum, joined to one another laterally by a pleural membrane.
64. The anus is nestled between three protective sclerites: a dorsal epiprocts and a pair of lateral paraprocts.
65. Median caudal filament: a thread-like projection arising from the center of the last abdominal segment (between the cerci).
66. Median caudal filament: found only in "primitive" orders (*e.g.* Diplura, Thysanura and Ephemeroptera).
67. Pincers: Dermaptera (earwigs); The cerci are heavily sclerotized and forceps-like.
68. Sting: modified ovipositor, found only in the females of (bees, and predatory wasps).
69. Homoptera: The rostrum is relatively short (1-3 segments) and emerges from near the ventral posterior margin of the head.
70. Heteroptera: The rostrum is relatively long (3-4 segments) and arises near the front or lower front of the head (prognathous or ahypognathous).
71. The ocelli are absent in butterfly.
72. Labial palp is snout like: Pyraustidae (grass borers).
73. In *Locusta migratoria* poststernal tubercal absent.
74. In *Schistocerca gregaria* poststernal tubercal present.
75. Johnson organ is found in the pedicel (the second segment of antenna). (ICAR-NET 2013).
76. Prehensile legs: Dragonflies- Catching prey (basket forming type).
77. Forewings are longitudinally divided twice forming a fork like structure whereas hindwings are divided twice in to three arms: Fissured wings.
78. The flagellar segment with no muscular attachment: (Annulated type).
79. The flagellar segment with muscular attachment: (Segmented type).
80. Clavate antenna with hook: skipper butterflies.
81. Mask type mouth part: Naids of dragonflies (labium modification).

Fill in the Blank Type Questions with Answers

1. **__________found on the head of larvae of Lepidoptera as light sensor.**

   Ans: Lateral ocelli

2. **Crochets are found on the prolegs of__________.**

   Ans: Lepidoptera

3. **Endocuticle and Exocuticle are jointly known as __________.**

   Ans: Procuticle

4. **__________is polymerized lipoprotein layer without chitin.**

   Ans: cuticulin

5. **__________is main component of cuticle.**

   Ans: Chitin

6. **Chitin is a polymer of__________(a nitrogenous polysaccharide).**

   Ans: N-acetylgulcosamine

7. **__________water insoluble tanned protin and gives colour to exocuticale.**

   Ans: Sclerotin

8. **__________is outer most layer of insect cuticle.**

   Ans: Cement

9. **Clothing hairs are present in__________insect.**

   Ans: Honey bee

10. **__________liner invaginations between two sclerites of the exoskeleton**

    Ans: Suture

11. **__________possesses 4 pairs of legs without crochets.**

    Ans: Mustard saw fly grub

12. **Head is connected to thorax by __________.**

    Ans: Cervix

13. **Pronotum saddle shaped in__________.**

    Ans: Grasshopper

14. **Spiracles are also called as __________.**

    Ans: Stigmata

15. **Abdominal segments are joined by intersegmental membrane known as __________.**

    Ans: Conjunctiva

16. __________is absent in abdominal segments.

 Ans: Pluron

17. Two lateral paraprocts and dorsal epiproct are the parts of __________abdominal segments.

 Ans: 11th

18. __________is well developed and triangular part of bugs.

 Ans: Scutellum

76. Order __________have distinct Y-shape epicranial suture.

 Ans: Pscoptera

77. Lacinia is rod like (pick) in order __________.

 Ans: Pscoptera

78. Mesothorax represented dorsally as Scutellum in __________.

 Ans: Hemiptera

79. Fringed wings present in __________insect.

 Ans: Thrips

80. Right mandible is absent in order __________.

 Ans: Thysanoptera

81. Halters are modified hind wings of order __________.

 Ans: Diptera

82. In honey bees, lapping tongue is formed by the__________ and __________.

 Ans: labium and maxillae

83. First abdominal segment of Hymenoptera is called__________.

 Ans: Propodeum

84. Except 1 and 2 abdominal segments, rest of the abdomen in hymenoptera is called as__________.

 Ans: Gaster or metasoma

Multiple Choice Questions

1. __________suture is known as ecdysial line which is inverted 'Y'shaped.

 (a) Epistomal (b) Epicranial

 (c) Clypeo-labral (d) Post occipital

2. Dorsal body sclerites of thorax are known as__________.

 (a) Sternum (b) Notum

 (c) Pleuron (d) Scutellum

3. **Hypognathus type of mouth parts found in__________.**
   (a) Grasshopper (b) Cockroach
   (c) Cricket (d) All of these

4. **Prognathus types of mouth parts are found in__________.**
   (a) Termites (b) Rhinoceros Beetles
   (c) Weevils (d) All of these

5. **In__________type of mouth parts head is deflexed and mouth parts are directed backword.**
   (a) Hemipteroid (b) Opisthgnathus
   (c) Opisthorhynchous (d) All of these

6. **__________type of mouth parts are found in *Nezara viridula* (JRF 2011).**
   (a) Opisthognathous (b) Prognathous
   (c) Hypognathous (d) None of these

7. **Basal segment of antennae is known as__________.**
   (a) Pedicel (b) Annuli
   (c) Clavola (d) Scape

8. **Antennae are innervated with nerves arise__________.**
   (a) Tritocerebrum (b) Protocerebrum
   (c) Deutocerebrum (d) All of them

9. **Johnston's organ present on __________segment of antennae.**
   (a) Pedicel (b) Scape
   (c) Fllagelum (d) All of these

10. **Pheromones are detected by __________.**
    (a) Maxillary palp (b) Antennae
    (c) Mandibulary palp (d) Both (a) and (b)

11. **Ant, weevils, wasp and honey bee have __________type of antennae.**
    (a) Flabellate (b) Pectinate
    (c) Geniculate (d) Stylet

12. **Cedar beetles have __________type of antennae.**
    (a) Flabellate (b) Serratet
    (c) Setaceous (d) Clavate

13. **Liquid feeding insects have __________type of mouth part.**
    (a) Mandibulate (b) Chewing
    (c) Lapping (d) Haustellate

14. ________type of teeth found in insect.

(a) Incisors (b) Molars

(c) Chisal like (d) Both (a) and (b)

15. Maxillary palp present on ________.

(a) Palpiger (b) Labium

(c) Palpifer (d) Mandible

16. Labial palp present on ________segment.

(a) Palpiger (b) Palpifer

(c) Labium (d) Mandible

17. Glossa and para-glossa fused together to form ________.

(a) Galae (b) Lapping tongue

(c) Ligula (d) Rostrum

18. Salivary gland duct open at the base of________.

(a) Epipharynx (b) Hypopharnyx

(c) Labrum (d) Mandible

19. Maxillary and labial palps are absent in________type of mouth parts.

(a) Sponging (b) Piercing and sucking

(c) Siphoning (d) Biting

20. Mandible and maxilla are modified into________in bugs.

(a) Glysets or fasciete. (b) Lapping tongue.

(c) Lingule (d) Rostram.

21. Mouthpart of mosquito have________ number of stylets.

(a) Four (b) Six

(c) Two (d) Five

22. Anticoagulants are present in saliva of ________.

(a) Mosquito (b) Housefly

(c) Fruitfly (d) Bug

23. Blunt and dumbel shaped mandibles without teeth found in________.

(a) Mosquito (b) Bug

(c) Honey bee (d) Housefly

24. Lapping tongue formed by the maxilolabial structure found in ________.

(a) Mosquito (b) Bug

(c) Honey bee (d) Housefly

25. **In Honey bee glossa ends into circular spoon shaped lobe know as__________.**

(a) Bouton (b) Flabellum

(c) Spoon (d) All of these

26. **Asymmetrical type of mouth parts found in__________.**

(a) Thrips (b) Mosquito

(c) Bug (d) All of these

27. **__________Mandible absent in thrips.**

(a) Left (b) Both 1 and 2

(c) Right (d) None of these

28. **Elongated Sickled shaped mandibles found in grub of__________insect.**

(a) White grub (b) Antlion

(c) Damselfly (d) Honey bee

29. **Psuedotrache found on labella of__________.**

(a) Housefly (b) Moth

(c) Butterfly (d) Honey bee

30. **Proboscis in Lepidoptera is formed by fusion of __________.**

(a) Galae of labium (b) Galae of Maxilla

(c) Labial pulp (d) Maxillaray pal

31. **A typical leg of insect have__________segment.**

(a) Five (b) Four

(c) Six (d) Seven

32. **Two segmented trouchanter found in __________.**

(a) Damselfly (b) Dragonfly

(c) Ichneuminid wasp (d) Both (a) and (b)

33. **__________is largest and short part of leg.**

(a) Femur (b) Trochanter

(c) Tibia (d) Coxa

34. **__________is a foot of insect.**

(a) Tarsus (b) Basitarsus

(c) Pretarsus (d) Tibia

35. **Empodium present in legs of __________.**

(a) Housefly (b) Honey bee

(c) Bed bug (d) Mosquito

36. Tibia works a thumb like process in__________ type of leg.

(a) Raptorial (b) Scansorial

(c) Foragial (d) Fossorial

37. Forelegs of honey bee have__________.

(a) Eye brush (b) Antenna cleaner or strigillis

(c) Pollen brush (d) All of these

38. Pollen basket or corbicula located at__________leg of Honey bee.

(a) Mid (b) Hind

(c) Fore (d) All three

39. Adult mite have__________ pair of legs.

(a) 4 (b) 2

(c) 3 (d) 1

40. __________muscles are directly connected with wings.

(a) Axillary muscle (b) Sub-alary muscle

(c) Lateral muscle (d) All of these

41. Wingless female and one pair of wings in case of male found in__________insect.

(a) Meloidae (b) Coccidae

(c) Saturnidae (d) All of these

42. Secondarilly wingless insect:__________

(a) Flea (b) Poultry louse

(c) Head louse (d) All of these

43. Decidious wings found in__________.

(a) Termites (b) Adult ants

(c) butterfly (d) flies

44. Insect wing is__________in shape.

(a) Semicircular (b) Circular

(c) Triangular (d) Rectangular

45. The angle between thorax and wing attachment is__________.

(a) Anal (b) Humeral

(c) Apical (d) Thoracic

46. The apical area of wing supported by vein is known as__________.

(a) Vanus (b) Jagum

(c) Remigium (d) None of these

47. Remigium and vannus areas of wing separated by__________.

(a) Vannal fold (b) Jugal fold

(c) Remigiual fold (d) Anal fold

48. __________is wing articulation sclerite.

(a) Axillary sclerite (b) Pteralia

(c) Axilla (d) None of these

49. __________type of wing have not veination and not used in flight.

(a) Elytra (b) Tegmina

(c) Scaly (d) Membranous

50. Fore wings of bugs are__________type.

(a) Elytra (b) Membranous

(c) Hemielytra (d) Halter

51. Halters are found in__________insects.

(a) Male strepsiptera and scale insect (b) Mosquito

(c) True flies (d) All of these

52. __________act as organ of balance in house fly.

(a) Fore wing (b) Scapel plate

(c) Elytra (d) Halters

53. Ovipositor of house fly is known as__________.

(a) Pseudovipositor (b) Ovitubus

(c) Oviscapt (d) All of these

54. In__________insect ovipositor is developed into sting.

(a) Dipteran (b) Worker honey bees

(c) Fruit fly (d) Housefly

55. Crickets' have__________like ovipositor.

(a) Needle (b) Sword

(c) Sting (d) Horny

56. Order orthoptera have__________wing.

(a) Hemielytra (b) Tegmina

(c) Elytra (d) Halters

Answer Key

1–b	11–c	21–b	31–a	41–b	51–d
2–b	12–a	22–a	32–d	42–d	52–d
3–d	13–d	23–c	33–a	43–a	53–d
4–d	14–d	24–c	34–c	44–c	54-b
5–d	15–a	25–d	35–a	45–b	55–a
6–a	16–b	26–a	36–b	46–c	56–b
7–d	17–c	27–c	37–a	47–b	
8–c	18–b	28–b	38–b	48–b	
9–a	19–b	29–a	39–a	49–a	
10–d	20–a	30–b	40–a	50–c	

Chapter 6

Anatomy, Biology and Physiology

1. Different layers of insect integument from outside: Epicuticle- Exocuticle- Mesocuticle- Endocuticle- Epidermis- Basement membrane.
2. Chitin is absent in epicuticle.
3. Different layers of epicuticle from outside: Cement layer- wax layer- polyphenol layer- cuticulin layer.
4. Precursor for ecdysone synthesis: Ergosteroid.
5. The concentration of juvenile hormone decreases as the larvae becomes pupae.
6. The alimentary canal is longer in phytophagous insects.
7. The expanded condition of the heart after diastole is called Diastasis.
8. Insect heamolyph is free from amino acids and with more trehalose.
9. Mullers organ: A group of numerous scolopophores forming a swelling applied to the inner surface of each tympanum and connected by the auditory nerve to the metathoracic ganglion.
10. Schwann cells or neurolemmocytes: the principal ganglia of the peripheral nervous system (PNS).
11. Hick's papillae: in Diptera; campaniform sensilla on the halter base.
12. Palmen's organ: tracheal system of mayfly.
13. Sempers rib: degenerated trachea in wings.
14. In protura the members of Eosentomoidea possess spiracles and a simple tracheal system, while those in the Acerentomoidea lack these structures and perform gas exchange by diffusion.

15. Cryptonephridial condition: most Coleoptera and Lepidoptera.
16. Peritropic membrane absent in hemiptera.
17. Palmitic and oleic acid prominent fatty acids in insect and comprises over 60 per cent.
18. Spiracles are absent (Apneustic condition) in chirnomid larvae.
19. RBC found in chirnomid larvae.
20. Malpighian tubules are absent in aphids.
21. Antennation: Touching with the antenna.
22. Teneral: Describing a condition in which the cuticle is soft and usually pale immediately after moulting.
23. Phagocytosis: The process of "cell eating" in which small particles or microorganisms are engulfed by blood cells.
24. The type of haemocytes which are responsible for the production of other type of haemocytes: proheamocytes.
25. Phagostimulant: A chemical that induces and maintains feeding.
26. Lipophorin: A plasma protein in insect hemolymph that transports lipids between tissues.
27. Oenocytes: (special cells associated with epidermis and tracheae) They are involved in synthesis of waxes, such as beeswax and waxes used for waterproofing cuticle and egg shells.
28. Peristigmatic glands: glands that secrete a hydrophobic material preventing wetting of the spiracles (in aquatic insect).
29. Fat bodies associated with storage excretion in insects.
30. Heamocoelous viviparity is prominent in the insect order: Strepsiptera.
31. Correct sequence: Spermatogonia- spermatocytes- spermatids- spermatozoa.
32. Carnitine is insect vitamin.
33. Alternation of generation is seen in: Aphids.
34. Adult diapause: white grub, *Anthonomus* and *Leptinotarsa.*
35. Extra carporeal digetion: when saliva is injected into the host, digetion starts before the food is injested.
36. The insect blood is usually green in colour due to the presence of Insectoverdin.

37. Insectoverdin is a mixture of blue and yellow compounds. The blue is usually mesobiliverdin, and the yellow is a carotenoid.
38. *Cutaneous respiration* is seen in many endoparasitic and aquatic *insects.*
39. Lines of weakness: (Anatomical feature) the suture lines where the integument will split to allow for moulting.
40. Meroistic ovary: An ovary possessing nurse cells that are connected to or accompany the oocytes.
41. Panoistic ovary: An ovary in which the ovarioles lack nurse cells.
42. Aphidilutein: a yellowish fluid found in plant lice, changed to a rich violet by alkaline reagents.
43. Cantharidin ($C_{10}H_{12}O_4$): the substance that gives the meloid beetles their blistering power composition.
44. Cuckoo spit: liquid in the form of bubbles produced by members of the family Cercopidae and which often conceals the producer.
45. Immature scales are called crawlers.
46. Cuckoo bees: lay their eggs in the nest cells of other bees.
47. All described genera of thrips are haplodiploid organisms capable of parthenogenesis.
48. Paedogenesis is common in cecidomyiidae.
49. Order having pupal stage only in male: Strepsiptera.
50. Fireflies are beetles that use light to attract a mate.
51. A particular Asian moth has evolved to feed on the tears of buffalo.
52. Trophallaxis: The exchange by oral transfer of alimentary liquid among members of a colony of social insects.
53. Nasute: A soldier in the family Termitidae that has a well-developed horn-like median projection (nasus) and reduced mandibles.
54. Dzierzon's rule: The original term for haplodiploid sex determination in honey bees, as defined by J. Dzierzon in 1845.
55. Lerp: sugary and waxy formations made by larval psyllids for their protection.
56. Neoteny: A condition in which nymphal or larval structures are carried over into the adult and the corresponding adult structures are suppressed.
57. Obligatory diapauses: A form of diapauses that is genetically programmed to occur at a specific developmental stage regardless of the environmental conditions that prevail.

58. Neosomy: The ability to produce new cuticle without moulting.
59. Neosome: An organism altered by neosomy.
60. Neosomule: A new external structure resulting from neosomy.
61. Biomorphotica: Those neuropterous insects in which the pupa is active.
62. Late adult Diapouse: *Danaus plexipus* (Monarch butterfly) and *Epilachna* (lady beetle).
63. Egg diapause: Grasshopper.
64. Larval diapause: Pink boll worm.
65. Prepupal diapouse: *Plodia interpuntella* (Indian meal Moth).
66. Pupal diapause: *Sarcophaga crassipalpis, S. argyrostoma* (flesh fly), *Manduca sexta* (Tobacco hornworm) and Red hairy caterpillar.
67. Embryo diapause (late embryonic): silk worm, and *Lymantrica dispar* (Gypsy moth).
68. Imaginal diapose: *Culex pipiens.*
69. Pupae are active in culicidae.
70. *Icerya pruchesi* is a bisexual insect.
71. Female genitalia are present in grasshopper on: 8th and 9th segment.
72. Male genitalia are found in grasshopper: 9th segment.
73. Stomodeum is covered with: six layers.
74. Four salivary glands are found in grasshopper.
75. The function of goblet cell: To produce digestive enzymes.
76. Blood vascular system in insect: Open type.
77. According to pH scale insect blood is acidic.
78. The albuminoid is present in insect blood: haemoprasin, haemochlorin and haemopyrhin.
79. The living cells that are floating in insect blood: haemocytes.
80. The male grasshopper has two testes.
81. The female grasshopper has two ovaries.
82. Sound producing and receiving organs are called as scopolophous organ.
83. Mciller's organ is a group of many sense organs, found in tympanum.

84. In grasshopper the tympanum organs is situated in 1st abdominal segment.
85. Nephrocytes are related with excretion.
86. Stemmata: visual organ in the larvae.
87. The respiration in mollusca takes place by Gills.
88. The expansion of termite queen's abdomen: Physogastry (ICAR-SRF, 2011).
89. The permanent thin film of air outside the body for gaseous exchange: plastron.
90. Only striated muscles are found in insects.
91. Acid secreated by cornicle of aphids: Myristic acid.
92. Bisexual insect: *Icerya purchesi.*
93. Immature form are capable of producing eggs or larvae: paedogenesis.
94. Rheotaxis: moves in a definite direction in relation to the flow of water.
95. Aerostats: a pair of large air sacs at base of abdomen in Diptera.
96. Androconia: specialized, usually small scales of peculiar form, found localized on some male butterflies.(ICAR SRF-2013).
97. Brin: the fluid silk thread from each salivary gland.
98. Colleterial gland: structure accessory to the oviduct, secreting the viscid material used in cementing the eggs together.
99. Neck or cervix is also known as crag.
100. Endotoky: form of reproduction where the eggs are developed within the body of the mother.
101. Entomogenous: growing in or on an insect.
102. Entomography: the description of an insect or of its life history.
103. Entomotaxy: the preservation and preparation of insects for study.
104. Hyperparasites are parasites (or parasitoids) of another parasite species.
105. Autoparasites are species in which the females feed on males to obtain a nutritional advantage.
106. Brood parasites are insects that live in the nests of social insects and feed on the juveniles.
107. Social parasites are insects that steal food or other resources from the nests of social insects.

108. Paedogenesis: reproduction in the sexually immature or larval stage.
109. A biological gill is an organ that allows dissolved oxygen from the water to pass (by diffusion).
110. Diffusion of gasses through the body wall: cuticular respiration (aquatic species).
111. Dragonflies differ from other aquatic insects by having internal gills associated with the rectum.
112. Tychopanthenogenesis: occasional or accidental parthenogenesis as seen in silkworm (*Bombax mori*).
113. Arrhenotoky: The unfertilised egg develops invariabily into a male, while the fertilised one into a female *e.g.* many hymenopterans (*Apis* spp.).
114. Thelytoky: The unfertilised egg develops into females *e.g.* Phasmids.
115. Deuterotoky: The unfertilised egg develops into both males and females *e.g.* Hymenoptera, *Oecophylla longinoda.*
116. The females are absent and those, appears as female have ovotestes: *Icerya pruchasii.*
117. Allantoin: a chemical secretion of the larvae that inhibits bacterial growth.
118. Glycerol, sorbitol: cryoprotective compound (prevent destruction of tissues so that insect can survive at freezing temperature) (ICAR-SRF-2011).
119. The apposition/photopic eyes are typical of diurnal insect.
120. Mosaic theory associated with insect vision.
121. Weismann's organ: cyclorrhapous dipterans- fusion of carpora allata, carpora cardiaca and prothorasic gland. (IARI Ph.D. 2011).
122. Olfactory receptors: detecters of odours (smells).
123. Gustatory receptors: tastes.
124. Cornicles-paired tube like structures, located dorsally on the abdomen of aphids.
125. Insects having open circulatory system.
126. Hypermetamorphosis is seen in the insect family: melloidae.
127. The dorsal diaphragm is formed by alary muscles of the heart and related structures; it separates the pericardial sinus from the perivisceral sinus.
128. The ventral diaphragm usually covers the nerve cord; it separates the perivisceral sinus from the perineural sinus.

129. Pulsatile organs are located near the base of the wings or legs.

130. About 90 per cent of insect heamolymph is plasma: a watery fluid, usually clear, but sometimes greenish or yellowish in colour. The remaining 10 per cent of heamolymph volume is made up of various cell types (heamocytes).

131. Most insects have only two types of visual pigments. One pigment absorbs green and yellow light (550 nm) and the other absorbs blue and blue and ultraviolet light (<480 nm).

132. Insects cannot see red colour.

133. Honeybees, bumblebees, and many diurnal butterflies have true colour vision.

134. True colour vision: three visual pigments with absorption maxima at 360 nm (ultraviolet), 440 nm (blue-violet), and 588 nm (yellow).

135. The true colour vision can also discriminate between single colors and mixtures of colors.

136. Ectodermal origin: Epidermis, exocrine glands, brain and nervous system, sense organs, foregut and hindgut, respiratory system, external genitalia.

137. Mesodermal origin: Heart, blood, circulatory system, muscles, endocrine glands, fat body, gonads (ovaries and testes).

138. Endodermal origin: Midgut.

139. Post-embryonic changes are known as morphogenesis.

140. The stylets with two maxillae and one mandible: Thysanoptera.

141. Factitious host: The host on which parasitoids oviposit readily, that they do not attack normally in the field. (IARI Ph. D. 2011).

142. The larvae of certain swallowtail butterflies have eversible glands, called osmeteria, located just behind the head (When a caterpillar is disturbed, it rears up, everts the osmeteria to release a repellent volatile).

143. Insect behaviour "play dead" (thanatosis): they release their grip on the substrate and fall to the ground where they are hard to find as long as they remain motionless.

144. Functional unit of excretion in earthworm: nephridia.

145. Pigmented cells are absent in superposition type of eyes.

146. Autotomy: some insects have a "fracture line" in each appendage (often between the trochanter and the femur) that allows a leg to break off easily if it is caught in the grasp of a predator.

147. Autotomy is common in crane flies, walkingsticks, grasshoppers, and other long-legged insects.
148. Locusts swarms at a speed of 12-16 km/hr.
149. Solitary phase of locust: the nymphs are green, gray or brown and the adults show shorter and saddle-shaped pronotum and shorter hind femur.
150. Gregaria phase of locust: The nymphs are black, yellow or orange and the adults show longer and crested pronotum and longer hind femur.
151. In many colonies of bees, ants, and wasps, worker females will remove eggs laid by other workers due to increased relatedness to direct siblings, a phenomenon known as worker policing.
152. Name related with 'Insect physiology': Wigglessworth.
153. Ants are closely associated with membracidae.
154. Only the insect that can moult in their winged form: mayfly (sub-imago).
155. Insect with five reproductive forms (Adult male, adult female, male producing larvae, female producing larvae, larvae producing both male and female): micromalthus (coleoptera).
156. Pupal forms that give birth to larvae: cecidomyids (*Tekamyia* and *Henria*).
157. Polyphenisum observed in: Aphids.
158. The unit of muscles between two z line: Sarcomere.
159. Pilifer of hawk moth: (unique auditory organ) present in head.
160. Polypod larvae: well segmentation (2-5 pair of legs on abdominal segment).
161. Looper larvae: legs on 6th and last segment.
162. Semilooper:legs reduced 3rd and 4th segment.

Fill in the Blank Type Questions with Answers

1. **Process of shedding of old cuticle with formation of new cuticle is known as__________.**

   Ans: Moulting

2. **At moulting, shedded cuticle is known as__________.**

   Ans: Exuviae

3. **__________ is separation of cuticle from epidermis at moulting.**

   Ans: Apolysis

4. __________is time between two moulting.

Ans: Stadium

5. __________is the form of insect in any stadium.

Ans: Instar

6. Arrangement of body segments into different regions is known as __________.

Ans: Tagmosis

7. Abdominal condition of queen termite is called as __________.

Ans: Physogastry.

8. Queen of termite is known as __________.

Ans: Egg laying machine.

9. Frontal glands are present in __________.

Ans: Nasute soldier

10. __________structure are present in termitoria of termites.

Ans: Fungal garden

11. __________type of pupa found in Hymenoptera.

Ans: Exarate

12. Males are developed from__________eggs and female from __________eggs in Hymenoptera, respectively.

Ans: Unfertilized and fertilized

13. Hymenopteran grub is __________.

Ans: Apodous

14. Hymenoptera male and female are__________ and __________, respectively.

Ans: Haploid and diploid

15. Polyembryony observed in __________.

Ans: Wasp

16. Larvae of mosquitoes are called as __________.

Ans: Wringler

17. Pupa of mosquitoes is called as __________.

Ans: Tumbler

18. Life cycle of__________pest is very short (7 days only).

Ans: Drasophilla

19. Eggs of Mallophaga are called as __________.

Ans: Nit.

20. **Physogastory is a condition observed in__________.**
    Ans: Termites

21. **Extra-oral digestion observed in order __________.**
    Ans: Hemiptera

22. **Naiad is aquatic stage found in order __________.**
    Ans: Odonata

23. **Secretion of honey dew is a common phenomenon of__________insects.**
    Ans: Homopteran

24. **Haemocoelic or traumatic insemination is phenomenon observed in __________insect.**
    Ans: Bed bug

25. **Pseudopupa observed in __________insect.**
    Ans: Whitefly

26. **Malphigian tubules are modified into silk glands in order __________.**
    Ans: Neuroptera

27. **Cornicals or siphonculi/wax tubes are present on __________abdominal segments of aphids.**
    Ans: 6th

28. **__________are swiftest among all insects.**
    Ans: Dipteran flies

29. **Dipterian pupa inclosed in last larval skin called __________.**
    Ans: Puparium

30. **In cockroach the ovipositor enclosed in a cavity formed by the enlarged seventh segment is known as__________.**
    Ans: Vestibulum

31. **Canabolism occurs in __________.**
    Ans: Mantids, Chrysoperla and Helicoverpa

32. **__________are called as true bugs.**
    Ans: Hetropterans

33. **__________have the repugnatorial or scent gland that emits an unpleasant odour.**
    Ans: Bugs

34. Honey dew secretion is common phenomenon of __________ suborder.

Ans: Homoptera

35. __________ have aquatic and quite active pupal stage

Ans: Mosquito

36. Haemoglobin is present in haemolymph of __________.

Ans: Chirnomids larvae.

Multiple Choice Questions

1. Foregut is also known as __________.

(a) Mesenteron (b) Stomodium

(c) Proctodium (d) Mesopteron

2. __________ act as sucking pump in sap feeder.

(a) Pharynx (b) Oesophagous

(c) Hypopharynx (d) None of these

3. __________ valve present between foregut and midgut.

(a) Pyloric (b) Cardiac

(c) Gizzard (d) None of these

4. Gizzard absent in __________ feeding insect.

(a) Leaf (b) Solid

(c) Sap or liquid (d) All of these

5. __________ teeth are present in gizzard of cockroach.

(a) Six (b) Seven

(c) Four (d) Five

6. Plate like internal cuticle of gizzard found in __________.

(a) Honey bee (b) Flea

(c) Bugs (d) Mosquito

7. __________ gut is endodermal in origin.

(a) Hind (b) Mid

(c) Fore (d) All of these

8. Peritrophic membrane present on internal lining of __________ gut.

(a) Hind (b) Fore

(c) Mid (d) All of these

9. Peritrophic membrane absent in__________feeding insect.

(a) Solid (b) Leaf

(c) Liquid (d) None of these

10. __________present between foregut and midgut.

(a) Gastric caecae (b) Hepatic caecae

(c) Enteric caecae (d) All of these

11. Midgut and hind gut are not connected in__________insect till pupation.

(a) Housefly (b) Honey bee

(c) Hoverfly (d) Robber fly

12. Absorption of food, water and salt takes place in__________gut.

(a) Mid (b) Fore

(c) Hind (d) None of these

13. Termites have pouch like__________to give place for housing symbionts.

(a) Ileum (b) Colon

(c) Rectum (d) Midgut

14. Saliva of honey bee have__________enzyme to digest sucrose.

(a) Invertase (b) Lipase

(c) Amylase (d) Protease

15. Extra intestinal digestion ocuures in__________insect, having pectinase enzyme to help in digestion.

(a) Moths (b) Bugs

(c) Butterfly (d) Borers

16. __________cells disintegrate at enzyme secretion process.

(a) Holocrine (b) Merocrine

(c) Perocrine (d) None of these

17. Gall midge saliva have__________chemical to produce galls in plant.

(a) IAA (b) IBA

(c) GA (d) All of these

18. Filter chamber present in__________.

(a) Fore-gut (b) Mid-gut

(c) Hind-gut (d) At the end of hind gut

19. __________act as organ of excretion.

(a) Malphigian tubule (b) Nephrocytes and oenocytes

(c) Fat bodies (d) All of these

20. __________cells of fat bodies responsible for storage and excretion.

(a) Trophocytes (b) Oenocytes

(c) Nephrocytes (d) None of these

21. Malphigian tubules are absent in __________.

(a) Aphid (b) Scale insect

(c) Cockroach (d) Locust

22. Malphigian tubules are present between__________.

(a) Stomodeum and Proctodeum (b) Fore gut and Hind gut

(c) Mesenteron and Proctodeum (d) All of these

23. __________is main organ of excretion and osmoregulation in insects.

(a) Nephrocytes (b) Rectum

(c) Malphigian tubule (d) None of these

24. __________substance excreted by terrestrial insect as a waste product.

(a) Ammonia (b) Urine

(c) Uric acid (Urates) (d) All of these

25. __________is excretory product in aquatic insects.

(a) Uric acid (b) Ammonia

(c) Urates (d) Na^+ and K^+

26. In cryptonephric condition __________membrane associated with rectum.

(a) Perinephric (b) Perotrophic

(c) Epidermal (d) Peritonic

27. Cryptonephric condition occurs in __________insect.

(a) Lepidopteros larvae (b) Coleopteran adult

(c) Larvae of Symphyta (d) All of these

28. Malphighian tubules act as organ of silk production in larvae of__________.

(a) Moth (b) Silkworm

(c) Neuroptera (d) Wasp

29. Uric acid stored in cells of__________in cockroach.

(a) Fat bodies (b) Mushroom gland

(c) Wings (d) Muscle

30. __________is waste product of pupal metabolism.

(a) Uric acid (b) Meconium

(c) Ammonia (d) Spittle

31. Caterpillar have__________type of respiratory system.

(a) Holopneustic (b) Peripneustic

(c) Hemipneustic (d) Metapneustic

32. When all spiracles are closed the respiratory system is known as__________.

(a) Metapneustic (b) Apneustic

(c) Parapneustic (d) None of these

33. Amphineustic type of respiratory system found in__________.

(a) Maggot (b) Wringler

(c) Moth (d) Bug

34. Which of the following not circulate in insect body?

(a) Antibodies (b) Hormones

(c) Nutrients (d) Protiens

35. Respiratory system having one pair of functional posterior spiracles called as__________.

(a) Amphipneustic (b) Metapneustic

(c) Propneustic (d) Monopneustic

36. Respiratory system having one pair of anterior spiracles called as__________, found in puparium.

(a) Metapneustic (b) Amphipneustic

(c) Propneustic (d) Monopneustic

37. __________is responsible for ring like appearance of tracheae.

(a) Taenidia (b) Cuticular lining

(c) Atrium (d) Peritrium

38. Through__________gaseous exchange occurs in insects.

(a) Tracheoles (b) Taenidia

(c) Atrium (d) Tracheoblast

39. Taenidia absent in__________.

(a) Honey bee (b) Cockroach

(c) Grasshopper (d) Cricket

40. Plastrons are hydrofuge hairs that hold film of air for respiration, found in__________.

(a) Water beetle (b) Water bug

(c) Maggot (d) Naiads

41. __________ type of circulatory system present in Insect.

(a) Open (b) Closed

(c) Both (a) and (b) (d) None of these

42. The dorsal vessel lies in__________sinus.

(a) Perineural (b) Pericardial

(c) Perivisceral (d) None of these

43. __________is pulsatile organ in insect circulatory system.

(a) Dorsal vessel (b) Aorta

(c) Ostia (d) Alary muscles

44. __________prevent the backflow of blood from dorsal vessel.

(a) Ostia (b) Cardial cell

(c) Aorta (d) Pyloric valve

45. Posterial part of dorsal vessel is called as__________.

(a) Heart (b) Aorta

(c) Ostia (d) Blood vessel

46. Cellular fraction of haemolymph is__________.

(a) Haemocytes (b) Trophocytes

(c) Oenocytes (d) All of these

47. Insect body mostly content__________sugar.

(a) Sucrose (b) Trehalose

(c) Glucose (d) Fructose

48. Haemolymph is __________in nature.

(a) Basic (b) Acidic

(c) Neutral (d) Highly basic

49. __________inorganic ion present in predator and parasitic insect in large amount.

(a) Na^+ (b) Mg^+

(c) K^+ (d) Cl^-

50. __________act as antifreezing compound in insect.

(a) Glycerol (b) Trehalose

(c) Haemolymph (d) Ammino acid

51. __________cells of haemolymph responsible for phygocytosis.

(a) Plasmatocytes (b) Prohaemocyte

(c) Cystocyte (d) Podocyte

52. __________are responsible for coagulation of blood and precipitation of plasma in insect.

(a) Cystocyte (b) Oenocytoids

(c) Plasmatocytes (d) Podocyte

53. __________are important neurotransmitting substance involved in impulse conduction.

(a) Acetylecholine (b) Dopamine

(c) Cholineestarase (d) Both (a) and (b)

54. Conduction of nerve impulses from sense organs to central nervous system done by__________.

(a) Sensory or afferent neuron (b) Motor or efferent neuron

(c) Inter neuron (d) All of these

55. __________ and __________ions responsible for axonal conduction of nerve impulses.

(a) Na^+ and Mg^+ (b) Na^+ and K^+

(c) K^+ and Cl^- (d) K^+ and Mg^+

56. Two ganglia are connected to each other by a transverse nerve fiber called as__________.

(a) Connectives (b) Commissures

(c) Ventral nerve cord (d) All of these

57. Brain is also known as__________ganglia.

(a) Thoracic (b) Sub Oesophageal

(c) Supra oesophageal (d) Oesophageal

58. __________control all insect body process and Coordination of different organs.

(a) Supra oesophageal ganglia (b) Sub oesophageal ganglia

(c) Para oesophageal ganglia (d) None of these

59. Compound eyes and ocelli are inervated by__________.

(a) Protocerebrum (b) Deutocerebrum

(c) Tritocerebrum (d) Supra oesophageal ganglia

60. An antenna is inervated by__________gangila.

(a) Protocerebrum (b) Deutocerebrum

(c) Tritocerebrum (d) Supra oesophageal ganglia

61. Sub oesophageal ganglia gives nerves to__________.

(a) Mandible (b) Maxilliae

(c) Labium (d) All of these

62. Genital organ and cerci inervated by__________ganglia.

(a) Thoraco abdominal (b) Thorasic

(c) Abdominal (d) All of these

63. __________visceral nervous system associated with aorta, brain and foregut.

(a) Stomato gastric (b) Ventral

(c) Caudal (d) Peripheral

64. __________nervous system associated with the receiving of stimuli from external environment by various sense organs.

(a) Peripheral (b) Visceral

(c) Central (d) Caudal visceral

65. Singal unit of compound eye is__________.

(a) Stemmata (b) Ommatidia

(c) Ocelli (d) Rhabdom

66. __________is light sensitive pigment in compound eye.

(a) Rhabdom (b) Rhodonin

(c) Rhodopsin (d) Lycopene

67. Image perception in holometabolous larvae is done by__________.

(a) Dorsal ocelli (b) Stemmata

(c) Stigmata (d) Antennae

68. Superposition type of ommitidia is found in__________insect.

(a) Diurnal (b) Nocturnal

(c) Both (a) and (b) (d) None of these

69. __________not involve in image seeing.

(a) Lateral ocelli (b) Dorsal ocelli

(c) Ommitidia (d) Compound eyes

70. __________attach ovary to insect body.

(a) Terminal filament (b) Ovarioles

(c) Calyx (d) Oviduct

71. Oocytes in various stages of development present in__________.

(a) Ovarioles (b) Terminal filament

(c) Lateral oviduct (d) Commone oviduct

72. __________is special nutritive cell for nourishment of developing oocyte in female.

(a) Phagocyte (b) Trophocyte

(c) Oenocyte (d) All of these.

73. Trophocyte or nurse cells absent in __________ovarioles.

(a) Panoistic (b) Telotrophic

(c) Polytrophic (d) Acrotrophic

74. Polytrophic ovarioles are found in__________.

(a) Moth and butterflies (b) Bugs

(c) Earwig (d) Whitefly

75. __________act as temporary storage of sperm in female.

(a) Spermatheca (b) Seminal vesicle

(c) Vagina (d) Oviduct

76. In __________viviparity larvae develop inside mother and full grown larva is deposited that undergo pupation immediately.

(a) Adenotrophic (b) Pseudoplacental

(c) Ovivivparity (d) Haemocoelus

77. Sporadic type of parthenogenesis observed in__________insect.

(a) Honey bee (b) Aphid

(c) Cockroach (d) Grasshopper

78. Thelytokous type of parthenogenesis observed in__________insect.

(a) Thrips (b) Wasp

(c) Midge fly (d) House fly

79. Arrhenotoky means the production of __________

(a) Female (b) Male

(c) Both (a) and (b) (d) None of these

80. Deuterotoky observed in__________insect.

(a) Cynipid wasp (b) Ear wig

(c) Fruitfly (d) Honey bee

81. Reproduction by means of immature stages of insect is known as __________.

(a) Paedogenisis (b) Polyembryony

(c) Parthenogenisis (d) Oviparity

82. Polyembryony was observed in __________insect.

(a) Wasp (b) Aphid

(c) Honey bee (d) Cockroach

83. Paedogenesis observed in ________insect.

(a) Gall midge (b) Thrips

(c) Wasp (d) Aphids

84. Hairy caterpillars have ________, that secrets irritant chemicals.

(a) Androconia (b) Scoli

(c) Stig glunol (d) All of these

85. Osmeteria are present in larvae of ________insect.

(a) *Papilio demoleus* (b) *Psyllepta derogota*

(c) *Euproctis subnotata* (d) *Aphis gossypii*

86. ________act as temporary storage and release of hormones secreted by neuroscretory cell.

(a) Corpora allata (b) Capora cardiaca

(c) Weismans ring (d) All of these

87. ________gland secretes juvenile hormone in insect.

(a) Carpora cardiaca (b) Salivary glands

(c) Carpora allata (d) All of these

88. ________is an antijuvenile hormone used as insecticide.

(a) Precocene (b) Rhodopsin

(c) Ecdysone (d) All of these

89. Weisman's ring found in________.

(a) Larvae of Lepidoptera (b) Nymph of Hemiptera

(c) Dipteran maggots (d) Pupa of coleoptera

90. Puparium hardening hormone in dipteran maggot secreted by________gland.

(a) Prothoracic gland (b) Weismann's ring

(c) Corpora allata (d) Neurosecretory cells

91. ________gland secretes moulting hormone ecdysone.

(a) Carpora allata (b) Prothoracic gland

(c) Carpora cardiaca (d) Salivary glands

92. Moulting continues during life of an insect, is________type of metamorphosis.

(a) Ametabola (b) Hemimetabola

(c) Holometabola (d) Hypermetabola

93. Pupal stage absent in________.

(a) Ametabola (b) Hemimetabola

(c) Pauromatabola (d) All of these

94. Larval form change at each instar in________.

(a) Hypermetamorphosis (b) Holometamorphosis

(c) Hemimetamorphosis (d) Parametamorphosis

95. Hypermetamorphosis found in________.

(a) Blister beetle (b) Ground beetle

(c) White grub (d) Jewel beetle

96. Nits are the eggs of________.

(a) Mosquito (b) Bed bug

(c) Head louse (d) Moths

97. Green lacewing have________type of eggs.

(a) Sullptured (b) Float

(c) Pedicellated (d) Egg raft

98. Caster butterfly have________type of eggs.

(a) Sculptured (b) Float

(c) Egg raft (d) Pedicellated

99. Culex mosquitos the egg are in________.

(a) Egg case (b) Egg raft

(c) Ootheca (d) Egg pod

100. ________lays an egg in mass covered with hairy scales known as disease free laying (DFL).

(a) Moths (b) Beetles

(c) Bugs (d) Weevils

101. Anal tuft of hair found on last abdominal segment of________.

(a) Rice stem borer (b) Rice moth

(c) Rice ear head bug (d) All of these

102. ________type of legs, only thoracic legs are present.

(a) Polypod (b) Oligopod

(c) Apodus (d) All of these

103. Body is 'C' shaped in________larvae.

(a) Campodiform (b) Erucifom

(c) Scarabiform (d) Slug caterpillar

104. Scoli are present on body of_________larvae.

(a) Slug caterpillar
(b) Citrus butterfly
(c) Caster semilooper
(d) Fruit borer

105. Campodeiform larvae found in_________.

(a) Lady bird beetle
(b) White grub
(c) Rhinocerous beetle
(d) All of these

106. Looper larvae have two pair of prolegs on_________abdominal segment.

(a) 6th and 11th
(b) 3rd and 6th
(c) 6th and 10th
(d) 2nd and 10th

107. Eucephalous type of larvae without leg found in_________.

(a) Red palm weevil
(b) Fruitfly
(c) Housefly
(d) Moths

108. Maggots of housefly is_________.

(a) Hemicephalous
(b) Acephalous
(c) Eucephalous
(d) All of these

109. When head capsule is absent in larvae is known as_________.

(a) Eucephalous
(b) Acephalous
(c) Hemicephalous
(d) None of these

110. When head capsule is well developed in larvae is known as_________.

(a) Eucephalous
(b) Semicephalous
(c) Acephalous
(d) cephalous

111. Chrysalis is an obtect pupa of_________.

(a) Butterfly
(b) Moths
(c) Mosquito
(d) Fruitfly

112. Obtect type of pupa found in_________.

(a) Red palm weevil
(b) Flies
(c) Moths
(d) Both (a) and (c)

113. When all appendages of body are free, the pupa is known as_________type.

(a) Obtect
(b) Tumbler
(c) Coarctate
(d) Exarate

114. Pupa of dipteran insect is_________type.

(a) Obtect
(b) Exarate
(c) Tumbler
(d) Coarctate

115. _________layer not permit the loss of water from body.

(a) Wax (b) Cuticulin

(c) Cement (d) Tectocuticle

116. Basal triangular segment of maxillae is_________.

(a) Stipes (b) Cardo

(c) Gulea (d) Lacinia

117. Labella of housefly marked by number of half opened small tubes known as_________.

(a) Trachae (b) Pseudotrachae

(c) Grooves (d) None of these

118. In grasshopper, tympanal organ present on_________abdominal segments.

(a) 1st (b) 2nd

(c) 3rd (d) 4th

119. _________is not organ of respiration.

(a) Trachae (b) Green glands

(c) Book lungs (d) Thorax

120. _________is true suture.

(a) Post-occipital (b) Epicranial

(c) Fronto genal suture (d) Frontocleapial

121. Wingless female and winged males are found in_________.

(a) Coccids (b) Strepsiptera

(c) Siphonoptera (d) Both (a) and (b)

122. Chrysalis is the pupa of_________

(a) Butterfly (b) Moth

(c) Beetle (d) Grasshopper

123. _________is precursor of ecdysone.

(a) Cholesterol (b) Cysteine

(c) Alkaloids (d) Ergosterol

124. Trachae originated from_________.

(a) Endoderm (b) Ectoderm

(c) Mesoderm (d) Cuticle

125. In holopneustic type of respiratory system_________spiracles are functional.

(a) 1 (b) 3

(c) 4 (d) All of these

126. Luciferin produced in__________organ.

(a) Hindgut (b) Malphigian tubules

(c) Foregut (d) Cerci

127. Storage excretion in insect done by__________.

(a) Fat bodies (b) Nephrocytes

(c) Lymphocytes (d) All of these

128. Nervous system of insect is__________.

(a) Chlorigenic (b) Andregenic

(c) Mesodermic (d) All of these

129. Reproductive organs are__________in origin.

(a) Ectoderm (b) Endoderm

(c) Mesoderm (d) None of these

130. Nervous system of insects originated from__________.

(a) Ectoderm (b) Mesoderm

(c) Endoderm (d) All of these

Answer Key

1–b	18–b	35–b	52–a	69–b	86–b
2–a	19–d	36–c	53–d	70–a	87–c
3–b	20–a	37–a	54–a	71–a	88–a
4–c	21–a	38–a	55–b	72–b	89–c
5–a	22–c	39–a	56–b	73–a	90–b
6–a	23–c	40-a	57–c	74–c	91–b
7–b	24–c	41–a	58–a	75–a	92–a
8–c	25–b	42–b	59–a	76–a	93-d
9–c	26–a	43–a	60–b	77–b	94–a
10–d	27–d	44–a	61–d	78–a	95–a
11–a	28–c	45–a	62–a	79–b	96–c
12–c	29–a	46–a	63–a	80–a	97–c
13–a	30–c	47–b	64–a	81–a	98–a
14–a	31–b	48–b	65–b	82–a	99–b
15–b	32–b	49-a	66–c	83–a	100–a
16–a	33–a	50–a	67–b	84–b	101–a
17–a	34–a	51–a	68–b	85–a	102–b

103–c	**108–b**	**113–d**	**118–a**	**123–a**	**128–a**
104–a	**109–b**	**114–d**	**119–d**	**124–b**	**129–c**
105–a	**110–a**	**115–a**	**120–a**	**125–d**	**130–a**
106–c	**111–a**	**116–b**	**121–d**	**126–b**	
107–a	**112–d**	**117–b**	**122–b**	**127–a**	

Chapter 7
Ecology

1. Reciprocal interactions over evolutionary time between phytophagous insects and their food plants, or between pollinating insects and the plants they pollinate, have been described as coevolution.
2. Coevolution: This term, coined by P. R. Ehrlich and P. H. Raven in 1964 from a study of butterflies and their host plants.
3. The study of plant galls is called cecidology, gall-causing animals (insects, mites, and nematodes) are cecidozoa, and galls induced by cecidozoa are referred to as zoocecidia.
4. Myrmecochory: seed dispersal by ants.
5. Domatia (little houses) may be hollow stems, tubers, swollen petioles, or thorns, which are used by ants either for feeding or as nest sites, or both.
6. Plants with true domatia, called ant plants or myrmecophytes,.
7. Rather than protection, some myrmecophytes derive mineral nutrients and nitrogen from ant-colony waste via absorption through the inner surfaces of the domatia. Such plant "feeding" by ants, called myrmecotrophy.
8. In the insect order odonata all nymphs are aquatic except *Megalagrion oahuense* which is terrestrial.
9. *Megalagrion oahuense*: species of damsel fly endemic to Hawaii.
10. The term ecology is derived from the Greek term "oikos" meaning "house" combined with "logy" meaning "the science of" or "the study of".
11. The term ecology was coined by a German biologist Earnst Haekel (1869).

12. Intraspecific competition: When members of population of the same species compete for resources we call it intraspecific competition.
13. Intraspecific competition: Cannibalism in American bollworm larvae.
14. Intraspecific competition: reduction in fecundity (egg laying) in rice weevil, *Sitophilus oryzae* during overcrowding.
15. Interspecific competition: This is the competition occurring between members of two or more species.
16. Gause's principle: The elimination of one species by another as a result of interspecific competition has come to be known as the competitive exclusion principle or Gause's principle.
17. Under high temperature and relative humidity conditions *Tribolium castaneum* eliminates *T. confusum* and vice versa under low temperature and RH conditions.
18. Subterranean nest: A nest below the surface of the ground.
19. Synovigenic: A form of reproduction in which an adult female continues to produce and to mature eggs throughout adult life.
20. Sun compass: The ability of an organism to use the sun to maintain a constant direction.
21. Euedaphon: An organism living wholly within the soil, often well below the surface and in the leaf litter.
22. Permissive host: A host insect species in which a parasite or pathogenic organism establishes a successful infection.
23. Myrmecophiles: Such organisms, generally insects other than ants, that live in or around ant nests and that exploit resources from the ants by begging, scavenging, or predation.
24. Life table: A tabulation of the life stages of an insect with a cumulative record of mortality and survival.
25. Light compass orientation: The orientation in which a constant angle with a light source (usually the sun) is maintained.
26. Phoresy: A condition in which an individual is carried about by another individual without harming that individual.
27. Scramble competition: Competition in which many individuals "scramble" for a limited resource, such as food.
28. Secondary defense: A defensive mechanism that is brought into play only in the presence of a threat-for example, an aggressive display.

29. Xylomycetophagy: inhabiting wood and consuming mycetes growing in wood or cultivating them for feeding.
30. Schisophagy: feeding on ground remains of plants and animals.
31. Saprophagy: feeding on decaying organic matter.
32. Sarconecrophagy: feeding on dead bodies of vertebrates.
33. Sapromycetophagy : inhabiting decaying matter and consuming mycetes growing inside or cultivating them for feeding.
34. Some insects are active in night: Nocturnal.
35. Some insect are active during the day: Diurnal.
36. Some insect active during dawn and dusk: Crepuscular.
37. The lowest temperature at which organisms can live indefinitely is called as minimum effective temperature.
38. Catastrophic factors: When the factors act in such way that constant proportion of population is destroyed irrespective of the number of individuals it contains.
39. Facultative factors: when the factors act in such a way which involve the destruction of a percentage of population with increase as the number of individuals per unit area increases.
40. Ecological backlash: The counter-response of pest population or other biotic factors in the environment that devalue the effectiveness of insect management.
41. Thermal constant: The total heat energy required to complete a certain stage of development in the life cycle of species is constant.
42. Insect which live in the places with neither excess nor deficiency of water: Mesocoles.
43. Speciation without geographic isolation: sympatric speciation.
44. Allopatric speciation or geographic speciation: speciation that occurs when biological populations of the same species become isolated from each other to an extent that prevents or interferes with genetic interchange.
45. Industrial melanism is observed in insect: *Biston betularia* (peppered moth).
46. Morris and Miller were first to adapt the life table format in the insect population of spruce bud worm (*Choristoneura fumiferana*).
47. Potential natality is the reproductive rate of the individuals in an optimal environment.

48. All insects are epimorphic (the number of body segments does not change over time).

49. The cornicles produce substances that repel predators or elicit care-giving behavior by symbiotic ants.

50. The petroleum fly (*Helaeomyia petrolei*), develops in pools of crude oil in the California oil fields.

51. The brine fly (*Ephydra cinera*), thrives in the high salinity of Utah's Great Salt Lake.

52. Insect pests with high reproductive rate and low survival rate are called r strategists.

53. Insect pest with high survival rate and low reproductive rate are called k strategists. (ICAR-SRF, 2011).

54. Any behaviour we can observe by watching an animal is overt behaviour.

55. Innate behaviour is genetically programmed. Individuals inherit a suite of behaviours (often called an ethogram).

56. Stereotypic: performed in the same way each time by each individual.

57. Learned behaviour: non-heritable, acquired only through observation or experience.

58. Instrumental learning depends on the animal's ability to remember the outcome of past events and modify future behaviour accordingly.

59. Latent learning involves memory of patterns or events when there is no apparent reward or punishment associated with the behaviour (learns the location of her nest site).

60. The most basic unit of innate behaviour is a simple reflex arc.

61. Orientation Behaviours are coordinated movements (walking, flying, swimming, etc.).

62. Kinesis is a change in the speed of movement (orthokinesis) or a change in the rate of turning (klinokinesis).

63. Taxis: movement directly toward (positive) or away from (negative) a stimulus.

64. Klinotaxis: involves side-to-side motions of the head or body with successive comparison of stimulus intensity as the animal moves forward.

65. Phototaxis: light; geotaxis: gravity; thigmotaxis: contact with other objects.

66. The dorsal light reaction is a special case (telotaxis) in which movement occurs at a constant 90° angle to a light source (moths tend to circle a street lamp at night).

67. The light compass reaction is another special case (menotaxis) in which insects (honey bees, for example) fly away from the nest site at a fixed angle (x°) to the sun and return at the supplementary angle (180-x°).

68. Bioaccessibility: The potential for a substance to come in contact with a living organism.

69. Bioavaibility: The potential for a substance to be taken up by a living organism.

70. Butterfly: Dianural in habit.

71. Moth: Nocturnal in habit.

72. Competitive exclusion principle/Gause's principle: The elimination of one species by another as a result of interspecific competition.

73. Batesian mimicry the mimic shares signals similar to the model, but does not have the attribute that makes it unprofitable to predators (*e.g.* unpalatability).

74. Batesian mimicry named after Henry Walter Bates.

75. Monarch (model) and the viceroy (mimic) butterflies are classical examples of Mullerian mimicry named after Fritz Muller. (IARI Ph. D. 2011).

76. Treehoppers (membracidae) mimic thorns to prevent predators from spotting them.

77. The ability to regulate the body temperature higher than the surrounding environment is called as endothermy.

78. Cricket chirping rate relates with surrounding temperature: Dolbear's low (ICAR-SRF 2013).

79. Cricket chirping rate and ant walking rate follows: Arrhenius equation (ICAR-SRF 2013).

Multiple Choice Questions

1. **The quantity of light of one foot candle falling over one sq. ft is called as__________.**

(a) Lumens (b) Hymen

(c) Microtom (d) Light ray

2. **__________ is the study of structure and function in nature.**

(a) Ecology (b) Physiology

(c) Ecosystem (d) Biome

(e) Carbaryl

3. **Insects' are__________ in nature.**
   (a) Thermic (b) Poikilothermic
   (c) None (d) Hydrophilic

4. **Ecology term first time proposed by__________scientist.**
   (a) Earnst Haeckel (b) Smith
   (c) Odum (d) Debach

5. **A self containing habitat in which living organisms and physical environment interact is called as__________.**
   (a) Biome (b) Ecosystem
   (c) Habitat (d) Niche

6. **__________ is a numerical aid used in the study of insect population.**
   (a) Chi square (b) RBD
   (c) CRD (d) Life table

7. **Light control locomotive activity of insects by direct action,this phenomenon is called as__________.**
   (a) Hygrokinesis (b) Phytokinesis
   (c) Photokinesis (d) Chemokinesis

8. **The intensity of light is expressed quantitatively in terms of__________.**
   (a) Lumens (b) Cumens
   (c) Pumens (d) Lacmens

9. **The movement of animal in response to light stimulus called a__________.**
   (a) Chemotataxis (b) Nechotatix
   (c) Phototaxis (d) Mechotatix

10. **The highest or potential of an insect to reproduce itself under ideal optimum condition called as__________.**
   (a) Rational potential (b) Biotic potential
   (c) Arational potential (d) Abiotic potential

Answer Key

1–a	3–b	5–b	7–c	9–c
2–c	4–a	6–d	8–a	10–b

Chapter 8
Crop Pest

1. Tussock Moths damage forests and shade trees by eating the leaves.
2. Gladiolus thrips: *Thrips simplex* (Morison).
3. Pest: derived from French word 'Peste' and Latin term 'Pestis' meaning plague or contagious disease.
4. A group of caterpillars are called an Army.
5. A group of cockroaches are called a Intrusion.
6. A group of crabs are called a Cast.
7. A group of lice are called a Flock.
8. A group of porcupines are called a Prickle.
9. A group of rats are called a Colony, Pack, Plague or Swarm.
10. Companionate planting: The intercropping of certain repellent plants with crop plants.
11. A serious pest of Mahua: *Inderbela* spp. (Bark Eating Caterpillars).
12. Cowbug (*Oxyrachis tarandus*) is pest of pulse crop.
13. Thrips infesting to pulse crop: *Taeniothrips* spp.
14. Only butterfly which infests pulse crop: Blue butterfly (*Lampides boeticus*).
15. The pest of pulse which pupates inside the pod: Red gram pod fly.
16. The bunch type variety of groundnut is susceptible to leaf minor.
17. Burnt appearance of groundnut fields is due to sever attack of: leaf minor (*Aproraema modiella*).

18. Digging of trenches around the groundnut fields: management of red hairy caterpillar.
19. Non insect pest of sunflower which feeds on the seeds: rose ringed parakeet (*Psittaculla krameri*).
20. Caterpillars feeds on the leaves of castor, leaving only midribs and veins: Slug (*Latoia lepida*).
21. Grub and adult bore into the top shoots of the cotton plant which leads to wilting of the plant: shoot weevil.
22. Cotton leaves turns into purplish red which fall off in severe cases: Cotton mite (*Tetranychus macfarlanei*).
23. Spotted bollworm: Small and circular hole with faecal matter on the fruiting bodies, Flared squares (Flare-up).
24. Pink bollworm: pin sized holes, rosette flowers.
25. The bacteria *Nematospora gossypii* stains the cotton fiber.
26. Stalk borer (*Chilo auricilius*) family- crambidae: Bores into the cane and feeds on the inner content.
27. Growing buds of mango shoots are attacked which results in formation of galls on leaf axil: Mango shoot gall marker (*Apsylla cistellata*).
28. The insect pest of mango which has quarantine importance: Mango stone weevil and mango fruit fly.
29. The pest of citrus where only the adults are damaging stage is citrus fruit sucking moth.
30. The pest of citrus where the early instars larva resembles bird dropping is citrus butterfly.
31. The larvae of citurs sucking moth feeds on weed hosts *viz., Tinospora cardifolia, Cocculus pendulus* and *Cocculus hirsutus.*
32. The bark eating caterpillar (*Indarbela tetraonis*) belongs to family Metarbelidae.
33. Mass trapping of adults of rhizome weevil using pheromone lures as Cosmolure and metalure.
34. Scientific name of banana rusty thrips: *Chaetonaphothrips signipennis.*
35. The ber fruit borer (*Meridarchis scyrodes*) belongs to family Carposinidae.
36. Rat tail like leaf tip with curling of chilli leaves is nature of damage: Chilli thrips.

37. Bore hole in the pseudo-stem and berries with accumulation of faecal pallets in cardamom (nature of damage): capsule borer.
38. Scientific name of spruce budworm: *Choristoneura fumiferna.*
39. Scientific name of turnip aphid: *Lipaphis erysimi.*
40. Cigarette beetle belongs to family: Anobiidae.
41. Virus transmission by mite: Circulative but non-propagative.
42. Virus transmission by mealy bug: semi-persistent.
43. Virus transmission by white fly: Circulative and persistent.
44. Virus transmission by thrips: Persistent and circulative.
45. Virus transmission by plant hopper: Circulative and propagative.
46. Scientific name of sesbania stem borer: *Azygophles sealaris* (Cossidae: Lepidoptera).
47. Scientific name of sesbania stem weevil: *Alcidodes bubo* (Curculionidae: Coleoptera).
48. Scientific name of daincha green semilooper: *Pericyma gloucians* (Noctidae: Lepidoptera).
49. Scientific name of daincha leaf webber: *Striglina scitaria* (Tryrdidae: Lepidoptera).
50. Scientific name of daincha green looper: *Semiothira pervolgata.*
51. Scientific name of daincha brown looper: *Hyposidra talaca successaria* (Geometridae: Lepidoptera).
52. Scientific name of sunhemp hairy caterpillar: *Utetheisia pulchella* Linn (Arctidae: Lepidoptera).
53. Scientific name of sunhemp mired bug: *Ragmus impartunitas* (Miridae: Hemiptera).
54. Gaint African snail (*Achatina fulica*) is controlled by predatory snail: *Gonaxis quadrilateralis.*
55. *Chrysopa* spp. is generally used as bioagent in IPM, but it is a pest which feeds on lac insect.
56. Nindyrin method is used for detection of hidden infestation in stored grains.
57. The term "Integrated Biological control" coined by Gurr and Wratten 1999.
58. Lepidopterocide: popular product by S. K. Mujumdar at CFTRI.

59. "TNAU-Agrobiocide": fusarium based product for controlling coconut mite.
60. Wet palm trapping done for rice thrips.
61. Council on Environmental Quality (CEQ, 1972) gave the term "Integrated Pest Management".
62. In 1997, Smith and Adkisson were awarded the World Food Prize for pioneering work on implementation of IPM.
63. Scientific name of green house snail: *Opeas gracilis.*
64. The scientific name of three stripped palm squirrel is *Funambulus palmarum* L. (ICAR-JRF 2009).
65. The scientific name of five stripped palm squirrel is *Funambulus pennanati.*
66. The scientific name of house sparrow is *Passer domesticus.* (ICAR-JRF 2009).
67. Cotton stainer: Dysdercus.
68. Early sowing and high seed rate in sorghum crop recommended for management of shoot fly. (ICAR-NET 2013).
69. Flowers are partially eaten in the pulse crop- nature of damage: Blister beetle.
70. Webbing of leaves, flowers and pods of pulses and feed by resting inside these webs - nature of damage: *Maruca testulalis.*
71. Seeds become shrivelled and lose germination in pulses- nature of damage: pod bug.
72. Ashy coat on apple stem: Apple wooly aphid (ICAR -NET 2012).
73. Monophagous pest: Yellow stem borer (ICAR-NET 2013).
74. Mango stone weevil: Monophagous pest (ICAR-NET 2013).
75. Skeletonization of cabbage leaves: DBM (ICAR -NET 2012).
76. Only nymphs are destructive: Psyllid.
77. 'Sliver shoot' in rice (nature of damage): *Orseolia oryzae.*
78. Common pest of neem trees: mealy scale (*Pulvinaria maxima*).
79. Appearance of rice field as grazed by cattle: cutworm (*Mythimna separata*).
80. Chemical responsible for elongation of leaf sheath (damage by gall fly): cecidogen.
81. Pseudo stem injection of banana, an injecting gun or hypodermic syringe is used for the control of banana aphid, vector of bunchy top disease.

82. Insect entered in India before enforcement of quarantine measures: Cottony cushion scale.
83. Amendment of Destructive Insect Pest Act was done in the year 1992.
84. Compulsory fumigation of imported cotton bales to prevent introduction of: Cotton boll weevil.
85. Insecticide Act of 1968 is replaced by 'Pesticide management bill-2008'.
86. Pest of Domestic quarantine importance: Cotton cushion scale, Coffee berry borer, San Jose scale, Codling moth.
87. Crow topping and eyummosis is damaging symptoms of red palm weevil.
88. "T" shaped marking on mango fruits is due to attack of *Sternochetus mangiferae.*
89. Pink colour encrustation on apple is due to damage of San Jose scale.
90. Insect act as a farmers friend at immature stage but harmful as adult: Blister beetle (ICAR-SRF, 2013).
91. The scientific name of prorcupine is *Hystrix indica*. (ICAR-JRF 2009).
92. *Euglandina rosea* is used for biological control of giant African snail.
93. Scientific name of neelgai is *Baseelaphus tragocamelus.*
94. Zinc phosphide is used in poison bait to control porcupine.
95. Katte disease of cardamom transmitted by aphid (*Pentalonia nigonervosa*).(ICAR-SRF 2011).
96. Black light trap is popular name for ultraviolet radiant energy with (320-380 nm wavelength).
97. Red ooze from sorghum spikelet when squeezed indicating the presence of maggot: sorghum midge.
98. *Phyllocnistis citrella* is a leaf miner pest of : *Citrus spp*. and other Rutaceae.
99. Webbing of flowers and pods in pigeon pea is due to *Maruca testulalis.*
100. The introduced pest in India: Coffee berry borer (ICAR-NET 2013).
101. Drying of cane with reduced internodes- nature of damage: Sugarcane scale (ICAR-NET 2013).
102. Mustard saw fly (pupation): soil (ICAR-NET 2013).
103. White grub (oviposition): loose soil (ICAR-NET 2013).
104. National Bureau of Agriculturally Important Insect (location): Bangalore (ICAR-NET 2013).

105. Spiral cut on the stem (nature of damage): Grapevine stem gridler (ICAR-NET 2013).
106. National centre for Integrated Pest Management (location): New Delhi.
107. Tomoto spotted wilt virus transmitted by: Thrips.
108. Yellow mosaic of bhendi transmitted by: white fly.
109. Pollu bettle: a pest of black paper (ICAR-NET 2013).
110. Citrus fruit sucking moth- nature of damage: shedding of citrus fruit with feeding punctures (ICAR-NET 2013).
111. Mechanical control by using iron hook: Rhinoceros beetle (ICAR-NET 2013).
112. The term ' Push and pull' was first conceved as astrategy for insect pest management (IPM) by Pyke *et al.* in Austalia 1967.
113. Pulse beetle is internal feeder. (ICAR-NET, 2012).
114. E.F. Knipling in 1937 in South East USA used the SIRM technique to control the screw wormfly (*Cochliomyia nominivorax*). (ICAR-NET 2013).
115. Cacao swellon shoot virus transmitted by mealy bug *Planococoides njalensis.*
116. Sterile insect technique (SIT): Knipling (ICAR-NET 2013).
117. Sterile insect technique (SIT) was successfullt employed in: Screw wormfly (*Cochliomyia hominivorax*). (ICAR-NET 2013).
118. Zero ETL used for the management of vectors.

Multiple Choice Questions

1. ___________is a common internal feeder on sorghum.

(a) *Sesamia inference* (b) *Chilo partellus*

(c) *Chilo infuscatelus* (d) *Earias vitella*

2. Grubs of blister beetle in the initial stage feed on the ___________

(a) Eggs of stem borer (b) Eggs of leaf hopper

(c) Eggs of red cotton bug (d) Eggs of grasshopper

3. *Epiricania melanoleuca* is the effective parasitoid of___________

(a) *Trichogramma* sp. (b) Rice stem borer

(c) Sugarcane pyrilla (d) *Earias vitella*

4. ___________is a fumigant type of rodenticide.

(a) Aluminium phosphide (b) Warfarin

(c) Zinc phosphide (d) Methyl bromide

5. **Drooping of top shoot in okra is due to attack of __________**
   (a) *Pectinophora gossypiella* (b) *Dysdercus cingulatus*
   (c) *Helicoverpa armigera* (d) *Earias vitella*

6. **The scientific name of American bollworm of cotton is__________.**
   (a) *Pectinophora gossypiella* (b) *Dysdercus cingulatus*
   (c) *Helicoverpa armigera* (d) *Earias vitella*

7. **__________pest oviposition is in soil around the stem.**
   (a) Aphids (b) Red cotton bug
   (c) Jassids (d) Pyrilla

8. **__________ is scientific name of white fly in Brinjal**
   (a) *Aleurocanthus woglumi* (b) *Aleurolobus barodensis*
   (c) *Bemisia tabbaci* (d) *Earias vitella*

9. **Scientific name of rice weevil is __________.**
   (a) *Trogoderma granarium* (b) *Sitophilus oryzae*
   (c) *Tribolium castanum* (d) *Latheticus oryzae*

10. **Carry over stage of sugarcane white grub is**
    (a) Egg (b) Nymph
    (c) Pupa (d) Adult

11. **Bunchy top in sugarcane is caused by __________.**
    (a) Stem borer (b) Top shoot borer
    (c) Root borer (d) Early shoot borer

12. **Staining of cotton lint is due to__________.**
    (a) Aphids (b) Dusky cotton bug
    (c) Root borer (d) Red cotton bug

13. **Give damaging stage of sesamum gall fly: __________**
    (a) Egg (b) Maggot
    (c) Nymph (d) Adult

14. **Damaging stage of ber fruit fly: __________**
    (a) Nymph and Adult (b) Maggot
    (c) Nymph (d) Grub and Adult

15. **The order for Mustard saw fly: __________**
    (a) Diptera (b) Hymenoptera
    (c) Hemiptera (d) Lepidoptera

16. In the system of Binomial nomenclature the first name indicates__________.

(a) Order (b) Family

(c) Genera (d) Species

17. In the system of Binomial nomenclature the second name indicates__________.

(a) Order (b) Family

(c) Genera (d) Species

18. 'IPM' stands for__________

(a) Integrated Pest Mechanism (b) Integrated Pest Management

(c) Important Pest Management (d) Important Pest Mechanism

19. The three names of the pest given below belongs to 'Borer complex' on cotton. Choose odd one.

(a) *Helicoverpa armigera* (b) *Earias vitella*

(c) *Chilo partellus* (d) *Pectinophora gossypiella*

20. The monophagous pest of the paddy: __________

(a) *Chillo partellus* (b) *Scirpophaga incertulas*

(c) *Nilparvta lugens* (d) *Nephotettix nigropictus*

21. 'Dead Heart' in paddy caused by__________.

(a) *Orseolia oryzae* (b) *Chilo partellus*

(c) *Nilparvata lugens* (d) *Nephotettix nigropictus*

22. Silver shoot/Onion leaf in rice caused by__________.

(a) *Scirpophaga incertulus* (b) *Chilo partellus*

(c) *Nilparvata lugens* (d) *Orseolia oryzae*

23. The order of Green leaf hopper is__________.

(a) Hemiptera (b) Homoptera

(c) Hymenoptera (d) Diptera

24. 'Hopper burn' symptoms in paddy is caused by__________.

(a) Green leaf hopper (b) Brown plant hopper

(c) Gall fly (d) Rice hispa

25. Buggy odour in rice field during milky stage is due to the pest: __________

(a) *Leptocorisa acuta* (b) *Nilparvata lugens*

(c) *Orseolia oryzae* (d) *Scirpophaga incertulus*

26. Appearance of paddy field as it grazed by cattle is due to the pest: ___________

(a) *Leptocorisa acuta* (b) *Nilparvata lugens*

(c) *Mythimna separata* (d) *Scirpophaga incertulus*

27. The important pest of nursery in paddy is___________.

(a) *Scirpophaga incertulus* (b) *Leptocorisa acuta*

(c) *Nephotettix nigropictus* (d) *Stenchaetothrips biformis*

28. Tungro virus disease complex is transmitted by___________.

(a) Green leaf hopper (GLH) (b) Brown plant hopper (BPH)

(c) Gundhi bug (d) Rice thrips

29. Passing the rope over paddy crop helps to manage___________.

(a) Paddy stem borer (b) Brown plant hopper

(c) Green plant hopper (d) Rice case worms

30. Clipping of paddy tips before transplanting is used for controlling ___________ pest.

(a) *Scirpophaga incertulus* (b) *Leptisa pygaea*

(c) Both (a) and (b) (d) None of these

31. The paddy stem borer pupates in the___________.

(a) Stem (b) On the stem

(c) In soil (d) All of the above

32. The family of yellow stem borer of paddy is___________.

(a) Noctuidae (b) Pyralidae

(c) Delphacidae (d) Cicadellidae

33. "Dead heart" in the sorghum is caused due to the pest/s: ___________

(a) *Atherigonia soccata* (b) *Chillo partellus*

(c) *Contarinia sorghicola* (d) Both (a) and (b)

34. Jowar shoot fly (*Atherigonia soccata*) pupates in/on:___________

(a) Soil (b) Stem

(c) Leaves (d) None

35. Midge fly (*Contarinia sorghicola*) pupates in: ___________

(a) Soil (b) Stem

(c) Spiklets (d) Leaves

36. Delphacids lay egg in/on___________.

(a) On stem (b) On leaves

(c) In soil (d) In leaf tissues

37. The sugary disease or Chikta is caused due to: ____________.

(a) *Atherigonia soccata* (b) *Contarinia soeghicola*

(c) *Peregrinus maidis* (d) *Mythmina separata*

38. The sorghum pest that attacks the crop when it is only up to four weeks is____________.

(a) *Atherigonia soccata* (b) *Contarinia sorghicola*

(c) *Chillo partillus* (d) *Perigrinus maidis*

39. The insect that infests sorghum crop exclusively after one month is____________.

(a) Sorghum shoot fly (b) Jowar stem borer

(c) Sorghum midge fly (d) Earhead bug

40. Workers caste in termites develops from the____________.

(a) Fertilized eggs (b) Unfertilized eggs

(c) Both (a) and (b) (d) None of these

41. There is____________queen in a colony.

(a) Only one (b) Two

(c) Three (d) Four

42. One pair of rat may give rise to about ____________young ones/year.

(a) 500 (b) 600

(c) 700 (d) 800

43. The poison bait of Zink phosphide recomanded for rat kill is of____________ per cent.

(a) 1 per cent (b) 2 per cent

(c) 3 per cent (d) 4 per cent

44. Zink phosphide is____________type of poison.

(a) Acute poison (b) Chronic poison

(c) Nerve poison (d) All of the above

45. ____________bait is available in the market as trade name 'Moosh moosh'.

(a) Zinc phosphide (b) Bromadiolone

(c) Warfarin (d) None of the above

46. The insect which feeds on silk and developing cobs in maize is____________

(a) *Helicoverpa armigera* (b) *Sesamia inferens*

(c) *Spodoptera litura* (d) *Earis vitella*

47. The chemical used for the control of rodents are called as___________.

(a) Acaricides (b) Miticides

(c) Rodenticides (d) Molluscides

48. The practice which consist of engaging person who digout rat burrows and kill the rats is___________.

(a) Trapping (b) Hunting

(c) Flooding (d) None of these

49. The commercial formulation of warfarin available in the market in name of ___________.

(a) Warin (b) Roden

(c) Rodafarin (d) Rodafarin'C'

50. Which acid is secreted by Blister beetle?

(a) Canthredin (b) Formic acid

(c) Salycyclic acid (d) Acetic acid

51. Give the family of *Mylabris postulate*: ___________

(a) Pyrilidae (b) Meloidae

(c) Noctuidae (d) Cerambicidae

52. The scientific name of Early stem borer of sugarcane crop is: ___________

(a) *Earis vitella* (b) *Chilo partellus*

(c) *Chilo infuscatellus* (d) *Sesamia inferens*

53. The scientific name of top shoot borer is: ___________

(a) *Earis vitella* (b) *Scirpophaga excerptalis*

(c) *Scirpophaga incertulus* (d) *Chilo infuscatellus*

54. Formation of galleries inside the cane tissue become red and cane become hard due to___________pets of sugarcane.

(a) *Chilo infuscutellus* (b) *Sesamia inferens*

(c) *Scirpophaga excerptalis* (d) *Chilo partellus*

55. The common internal feeder infesting sugarcane, maize and wheat is___________.

(a) *Sesamia inferens* (b) *Chilo partellus*

(c) *Athalia proxima* (d) *Chillo infuscatellus*

56. 'Bunchy top of sugarcane' is observed due to___________pest of sugarcane.

(a) Early shoot borer (b) Top shoot borer

(c) Internode borer (d) Root borer

57. Burning of sugarcane trash is done to prevent__________pest in succeding crop.

(a) Early shoot borer (b) Top shoot borer
(c) Internode borer (d) Root borer

58. Give the family of *Pyrilla purpusilla*: __________

(a) Aphididae (b) Delphacidae
(c) Fulgoridae (d) Pyralidae

59. The lepidopterous nymphal parasitoid used to control the *Pyrilla purpusilla* is__________.

(a) *Epiricania melanoluca* (b) *Apentalis glomeratus*
(c) *Trichograma chilonis* (d) *Chilonus blackburni*

60. The scientific name of sugarcane wooly aphids is__________.

(a) *Aphis crassivora* (b) *Aphis sacchari*
(c) *Ceratovacuna lanigera* (d) *Myzus persicae*

61. Tobbaco leaf eating caterpillar pupates in__________.

(a) Soil (b) On leaves
(c) Silken cocoons (d) Earthen cocoons

62. The scientific name of tea mosquito bug is __________.

(a) *Helopeltis antonii* (b) *Ferrisia virgata*
(c) *Inderbela tetraonis* (d) *Nezara virudala*

63. Family of tea mosquito bug is

(a) Pentatomidae (b) Pseudococcidae
(c) Miridae (d) Pyralidae

64. Family of coffee seed borer is__________

(a) Scolytidae (b) Pyralidae
(c) Miridae (d) Noctudae

65. Damaging stage of coffee seed borer: __________

(a) Grub (b) Beetle
(c) Larva (d) Nymph

66. Damaging stage of rhizome fly: __________

(a) Nymph (b) Adult
(c) Maggots (d) Grub

67. Young ones of thrips called as ___________.

(a) Nymph (b) Adult

(c) Maggots (d) Grub

68. Pollu beetle of black paper is also known as___________.

(a) Floes beetle (b) Lea beetle

(c) Blister beetle (d) Flat beetle

69. Scientific name of pollu beetle is___________.

(a) *Holotrachia serrata* (b) *Longitarsus nigripenis*

(c) *Mylabris pustulata* (d) *Athelia proxima*

70. The nymphs of tea mosquito bug are parasitised by___________nematode.

(a) *Steinernema carpocapsi* (b) *Meloidogyne incognita*

(c) *Agamermis paradecaudate* (d) *Rotylenchulus reniformis*

71. Mealy bug of black paper is primarily pest of___________crop in green house.

(a) Anthurium (b) Chrysanthemum

(c) Rose (d) Tuberose

72. The order of rose white fly is___________.

(a) Diptera (b) Hymenoptera

(c) Hemiptera (d) Heteroptera

73. Damaging stage of pollu beetle is___________.

(a) Grub (b) Adult

(c) Grub and Adult (d) Nymph

74. Damaging stage of fig jassid is___________.

(a) Nymph (b) Adult

(c) Nymph and adult (d) Only Grub

75. Scientific name of fig jassid is___________.

(a) *Amrasca bigutula bigutula* (b) *Amritodus atkinsoni*

(c) *Velu caricae* (d) *Empoasca kerri*

76. Armyworm, *Mythimna separata* has___________habit.

(a) Diurnal (b) Nocturnal

(c) Diurnal and Nocturnal (d) Dawn

77. Oviposition of white grub in___________.

(a) In soil (b) In earthen cells

(c) On leaves (d) On stem of plant

78. The carryover stage in white grub is__________.

(a) Beetle (b) Eggs

(c) Grub (d) Pupa

79. Workers of termites feeds on: __________

(a) Only stem (b) Only root

(c) Root and Stem (d) Leaf

80. The family of sugarcane woolly aphid is__________.

(a) Pempsigidae (b) Aphididae

(c) Delphacidae (d) Fulgoridae

81. The larval parasitoid of sugarcane woolly aphid is__________.

(a) *Trichogramma chilonis* (b) *Crysoperla carnea*

(c) *Cryptolaemus mountrouziari* (d) *Dipha aphidivora*

82. The scientific name of sugarcane mealy bug is__________.

(a) *Maconellicoccus hirsute* (b) *Bemisia tabaci*

(c) *Aleurocanthus woglumi* (d) *Saccharicoccus sacchari*

83. The scientific name of sugarcane whitefly is__________.

(a) *Bemisia tabaci* (b) *Aleurolobus barodensis*

(c) *Aleurocanthus woglumi* (d) *Ceratovacuna lanigera*

84. Clipping and destroying sugarcane leaves is done to control__________ pest.

(a) Sugarcane wooly aphids (b) Early shoot borer

(c) Sugarcane white fly (d) Sugarcane pyrilla

85. The vector of 'Moulting and spike' disease of sugarcane is__________.

(a) Sugarcane wooly aphid (b) Sugarcane white fly

(c) Sugarcane pyrilla (d) Sugarcane mealy bug

86. Name of the fungus which developes (black sooty mould) on the honey dew like secretion insects: __________

(a) *Aspergillus* spp. (b) *Trichoderma* spp.

(c) *Capnodium* spp. (d) *Colletotritum* spp.

87. 'Rosette flowers' in cotton is due to__________pest.

(a) *Helicoverpa armigera* (b) *Earias vitella*

(c) *Earias insulaha* (d) *Pectinophora gossypiella*

88. Flared square (Flare up) in cotton is due to___________pest.

(a) *Earias insulana* (b) *Helicoverpa armigera*

(c) *Pectinophora gossypiella* (d) All of the above

89. Dropping and drying of shoots in cotton crop is due to___________pest.

(a) *Earis insulaha* (b) *Helicoverpa armigera*

(c) *Pectinophora gossypiella* (d) All of the above

90. Where the cotton jassid (Leaf hopper) lay eggs?

(a) In the leaf tissue (b) On the tender parts of leaves

(c) Lower side of leaves (d) Upper side of leaves

91. Scientific name of Groundnut leaf minor is___________.

(a) *Aproaerema modicella* (b) *Aphis crassivora*

(c) *Chromatomyia horticola* (d) *Elasmolemus sordidus*

92. Family of Groundnut leaf minor is___________.

(a) Arctiidae (b) Gelechidae

(c) Noctuidae (d) Pyralidae

93. Family of Red hairy caterpillar is___________

(a) Gelechidae (b) Noctuidae

(c) Pyralidae (d) Arctiidae

94. The pest responsible for transmitting "Tomato spotted wilt virus" and for "Bud necrosis" disease is___________.

(a) Red hairy caterpillar (b) Groundnut leaf minor

(c) Groundnut thrips (d) Groundnut aphid

95. The scientific name of pulse beetle is___________.

(a) *Sitophylus oryzae* (b) *Tribolium castaneum*

(c) *Callasobruchus chinesis* (d) *Rhizopertha dominica*

96. The family of pulse beetle is ___________.

(a) Curculionidae (b) Bruchidae

(c) Bostrychidae (d) Gelechidae

97. The damaging stage of pulse beetle is ___________.

(a) Grubs (b) Beetles

(c) Both Grubs and beetles (d) Caterpillar

98. The family of rice weevil is___________.

(a) Bruchidae (b) Curculionidae

(c) Bostrychidae (d) Apionidae

99. The scientific name of red rust flour beetle is__________

(a) *Callasobruchus chinesis*
(b) *Rhizopertha dominica*
(c) *Sitophilus oryzae*
(d) *Tribolium castenium*

100. Which of the following store grain pest eggs are used for the mass multiplication of *Trichograma chilonis*?

(a) *Sitophilus oryzae*
(b) *Corcyra cephalonica*
(c) *Rhizopertha dominica*
(d) *Cylas formicarius*

101. When any pest occurs frequently on crop is known as________.

(a) Regular pest
(b) Persistent pest
(c) Endemic pest
(d) Epidemic pest

102. When any pest occurs on crop throughout the year is called as ________.

(a) Regular pest
(b) Persistent pest
(c) Endemic pest
(d) Epidemic pest

103. The infestation of pest is a regular phenomenon and occurs in some isolated localities is called ________pest.

(a) Regular pest
(b) Persistent pest
(c) Endemic pest
(d) Epidemic pest

104. When the infestation of pest in severe form in particular localities at particular season is known as________pest.

(a) Regular pest
(b) Persistent pest
(c) Endemic pest
(d) Epidemic pest

105. Which of the following pest are introduced in India from foreign countries?

(a) *Phthorimaea opercullella*
(b) *Coccus viridis*
(c) *Plutella xylostella*
(d) All of these

106. Which one of the following pests of guava introduced in India ?

(a) *Phthorimaea opercullella*
(b) *Coccus viridis*
(c) *Aleurodicus dispersus*
(d) None of these

107. Which of the following is monophagous pest of paddy?

(a) *Scirpophaga incertulas*
(b) *Sesamia inference*
(c) *Chilo partellus*
(d) *Chilo infuscatullus*

108. 'Dead heart' in paddy produced by__________.

(a) *Scirpophaga incertulas*
(b) *Chilo infuscatullus*
(c) *Chilo supressalis*
(d) *Emmalocera depressella*

109. Clipping of leaf tips of paddy during transplanting done to control________ pest.

(a) Rice case worm (b) Paddy gall midge

(c) Rice skipper (d) Paddy stem borer

110. Paddy seedling roots are dipped in________insecticide to control stem borer.

(a) Quinolphos (b) Phorate

(c) Imidacloprid (d) Phosphomidon

111. ________is used as biological control for *Scirpophaga incertulas*.

(a) *Trichogramma japonicum* (b) *Trichogramma chilonis*

(c) *Chrysoperla carnae* (d) *Cryptolaemus montrozieri*

112. Silver shoot emergence in paddy is characteristic damage of________ pest.

(a) *Chillo partellus* (b) *Scirpophaga incertulas*

(c) *Orseolia oryzae* (d) *Nephotettix virescens*

113. *Nilparvata lugens* produces________type of symptoms in paddy.

(a) Hopper burn (b) Yellowing

(c) Greening (d) Galleries of leaf

114. Lodging of paddy crop observed by________infestation.

(a) Green leaf hopper (b) BPH

(c) Gall midge (d) All of these

115. ________is known as swarming caterpillar.

(a) *Spodoptera mauritia* (b) *Helicoverpa armigera*

(c) *Mythimna separata* (d) *Melanitis ismene*

116. Irregular white patches on rice leaf caused by________.

(a) *Spodoptera litura* (b) *Nephotettix virescense*

(c) *Dicladispa armigera* (d) *Hieroglyphus banian*

117. Scientific name of paddy grass hopper is ________.

(a) *Chrotogonus* spp. (b) *Hieroglyphus banian*

(c) *Colemenia spheneriodes* (d) *Pulvinaria maxima*

118. Linear white patcthes along the leaf vein in paddy caused due to________pest.

(a) Paddy blue beetle (b) Leaf hopper

(c) BPH (d) Rice hispa

119. The scientific name of rice hispa is________.

(a) *Dicladispa armigera* (b) *Leptisma pygmaea*

(c) *Cnaphalocrosis medinalis* (d) *Melanitis ismene*

120. Tubular cases formation in paddy due to________.

(a) Paddy leaf roller (b) Rice case worm

(c) Rice butterfly (d) Paddy leaf folder

121. Rope dragging in paddy is done to control________ pest.

(a) *Amasacta moorei* (b) *Cnaphalocrosis medinalis*

(c) *Nymphula depunctalis* (d) *Nilparvata lugens*

122. Family of *Pelopidas mathias* is________.

(a) Coreidae (b) Miridae

(c) Satyariidae (d) Hesperiidae

123. ________pest attack on paddy at milky stage.

(a) *Leptocorisa acuta* (b) *Scirpophaga incertulus*

(c) *Nymphula depunctalis* (d) *Cnaphalocrosis medinalis*

124. Trap cropping is________control method.

(a) Mechnical (b) Biological

(c) Cultural (d) Physical

125. Madras Agricultural Pest and Disease Act enacted in________year.

(a) 1920 (b) 1914

(c) 1919 (d) 1918

126. Insecticide Act comes into enforced on________.

(a) 2nd Sept. 1968 (b) 2nd Sept. 1978

(c) 1st Jan. 1968 (d) 1st Jan. 1971

127. Federal Environmental Pesticide Control Act (FEPCA) came in________year.

(a) 1947 (b) 1972

(c) 1971 (d) 1948

128. Use of light traps, is________control method.

(a) Mechanical (b) Biological

(c) Physical (d) Chemical

129. Cottony cushion scale was controlled by________.

(a) *Mantis religiosa* (b) *Rodolia cardinalis*

(c) *Cryptolaemaus montrouzieri* (d) None of these

130. *Opisina arenosella* was controlled by________larval parasitoid.

(a) *Trichospilus pupivora* (b) *Trichogramma chilonis*

(c) *Goniozus nephantidis* (d) *Nephantis serinopa*

131. 'Indian station of Commonwealth institute of Biological control' located at_________.

(a) New Delhi (b) Hyderabad

(c) Mumbai (d) Banglore

132. *Paracocus marginatus* controlled by_________parasitoid.

(a) *Trichogramma* spp. (b) *Acerophagus papayii*

(c) *Goniozus nephantidis* (d) *Chrysoperla carnae*

133. DD-136 complex discovered by_________in 1955.

(a) Knipling (b) Dasture

(c) Dutky and Hough (d) Smith

134. DD-136 complex have nematode *Neoplectana carpocapsae* and bacteria _________.

(a) *Bacillus papillae* (b) *Corhenobacter tritici*

(c) *Achromobacter nematophilus* (d) *Bacillus thuringenisis*

135. Dead body of insect larvae is known as_________.

(a) Dead larvae (b) Callus

(c) Cadavour (d) None of these

136. Spore forming crystalliferous bacteria, *Bacillus thuringiensis* was isolated by _________.

(a) Smith, 1919 (b) Dutky and Hough, 1955

(c) Berliner, 1915 (d) Zeidler, 1912

137. *Bacillus thuringiensis* is _________poison.

(a) Contact (b) Systemic

(c) Stomach (d) Non of these

138. *Bacillus thuringiensis* act in on_________gut pH.

(a) Acidic (b) Alkaline

(c) Neutral (d) Slightly alkailine

139. *Bacillus thuringenisis* is mostly toxic to the larvae of _________order.

(a) Coleoptera (b) Lepidoptera

(c) Hymenoptera (d) Diptera

140. Spore count in commercial formulations of Bt is _________.

(a) 30-40 billion spore/g (b) 60-70 billion spore/g

(c) 90-102 billion spore/g (d) 10-20 billion spore/g

141. *Bt* was first isolated by Berliner in 1915 from the _________ insect.

(a) Wax moth
(b) Mediterranean flour moth
(c) Codling moth
(d) Sphinged moth

142. *Bt* var. *israelensis*, serotype -H-14 is used to control larvae of _________.

(a) White grub
(b) Mosquito
(c) Fruit fly
(d) Rice moth

143. Which of the following fungus used to control *Pyrilla purpusilla*?

(a) *Entomopthora grylli*
(b) *Entomopthora lecanii*
(c) *Meterrhizum anisopliae*
(d) *Verticillium* spp.

144. _________ fungus used to control most of homopteran sucking pest like white fly, aphid and scales

(a) *Beauveria bassiana*
(b) *Entomopthora grylli*
(c) *Verticillium lecanii*
(d) None of these

145. Commercial cultivation of Bt-cotton was started in India in _________ year.

(a) 2001
(b) 2006
(c) 2002
(d) 1996

146. Which of the following fungus used to control cabbage butterfly?

(a) *Metarrhizium anisopliae*
(b) *Neummoria spp.*
(c) *Entomophthora sphaerosperma*
(d) *Aspergillus flavous*

147. The toxins dexstruxins- A, B, C, D, E and F are produced by _________ fungus.

(a) *Aspergillus flavous*
(b) *Beauveria bassiana*
(c) *Metarrhizium anisopliae*
(d) All of these

148. *Beauveria bassiana* produces _________ toxins that kills the insect pest.

(a) Beauverolides
(b) Beauvericin
(c) Bassinolide
(d) All of these

149. Cry 1 Aa gene of Bt is specific to _________ pest.

(a) Lepidopteran
(b) Dipteran
(c) Coleopteran
(d) All of these

150. Cry 3 A gene of Bt is specific to _________ pest.

(a) Lepidopteran
(b) Coleopteran
(c) Both (a) and (b)
(d) None of these

151. _________ pest developed the resistance to Bt cotton having Cry 1 Ac gene.

(a) Pink bollworm
(b) Pink borer of maize
(c) Spotted bollworm
(d) None of these

152. The plant that can withstand the attack of insect pest and gives optimum yield is known as_________plant.

(a) Tolerant (b) Resistant

(c) Antibiosis (d) Hybrid

153. The resistance of plant to certain pest due to presence of some toxic chemical or due to lack of nutritional factor is known as_________.

(a) Antibiosis (b) Antixenosis

(c) Tolerance (d) Vertical Resistance

154. The resistance of plant to certain pest due to presence of specific morphological character like colour, hairyness, aroma etc. is known as_________.

(a) Antibiosis (b) Antixenosis

(c) Tolerance (d) Horizontal resistance

155. Application of sticky band around the trees to control__________pest.

(a) Whiteflies (b) Mealy bugs

(c) Scale (d) Bugs

156. Apple wooly aphid controlled by__________parasitoid.

(a) Mites (b) *Aphelinus mali*

(c) Both (d) None of these

157. Fruit fly controlled by using____________.

(a) Protien hydrolysate (b) Methyl eugenol

(c) Yeast (d) Both (a) and (b)

158. Which of the following is legume aphid?

(a) *Apis dorsala* (b) *Aphis* sp.

(c) *Aphis craccivora* (d) All of these

159. __________crop is used as a trap crop in cotton.

(a) Tomato (b) Cabbage

(c) Mustard (d) Okra

160. Mustard is used as trap crop in cabbage to attract__________pest.

(a) Black moth (b) DBM

(c) Hairy catterpiller (d) Semi looper

161. Clipping of tips of rice seedling is__________method of pest control.

(a) Physical (b) Cultural

(c) Mechanical (d) Biological

162. Rope dragging in rice field used to control__________pest of paddy.

(a) Stem borer (b) Aphids

(c) Case worm (d) BPH

163. Tin bands are fixed over the coconut plants to prevent the__________pest of coconut.

(a) Rat (b) Rhinoceros beetle

(c) Mealy bug (d) Mites

164. Alkathane bands cover around the trunk of mango to prevent the__________pest of mango.

(a) Rat (b) Mealy bug

(c) Jassids (d) Mites

165. Heating of empty godowns to a temperature above 50 °C for 10 to 12 hours is__________method of pest control.

(a) Cultural (b) Physical

(c) Chemical (d) Biological

166. The gamma rays are produced by the isotypes like__________to control the store grain pest.

(a) Co^{40} (b) Co^{50}

(c) Co^{30} (d) Co^{60}

167. Beetles complete their life cycle by passing __________ stage.

(a) Egg, Nymph (b) Egg, Nymph, Adult

(c) Egg, larva, Nymph (d) Egg, larva pupa, adult

168. In storage grain moisture percentage should be __________ so, pest will not infest grain.

(a) 5 per cent (b) 7 per cent

(c) 9 per cent (d) 12 per cent

169. Use of light trap is which type of pest control method?

(a) Mechanical (b) Physical

(c) Chemical (d) Cultural

170. Early earthing up is which type of pest control method?

(a) Physical (b) Mechanical

(c) Cultural (d) Chemical

171. National Center for IPM located at

(a) Mumbai (b) New Delhi

(c) Bangalore (d) Pune

172. __________ per cent humidity is considered as favorable for the normal development of insect.

(a) 60-70 (b) 40-50

(c) 80-90 (d) 100

173. Combination of all possible insect control methods in compitable way is called __________.

(a) Legislative method (b) Physical method

(c) Biological method (d) IPM

174. __________crop is used as trap crop in cabbage field.

(a) Tur (b) Bhendi

(c) Mustard (d) Til

175. Locust control board was located in__________state.

(a) Delhi (b) Maharashtra

(c) Rajasthan (d) M. P.

176. The average population density of pest over a long period of time unaffected temporary intervention of pest control is called as__________.

(a) ETL (b) GEP

(c) EIL (d) RL

177. __________refers to supervision of incidence of population of pest once in three or five years.

(a) Survillance (b) Rovar Survey

(c) Plane survey (d) Survey

178. Constant watch on population dynamics of the pest is called as__________.

(a) Constant survey (b) Survillance

(c) Rovar survey (d) Survey

179. The pest occure frequently and have close relationship with crop is called as__________pest.

(a) Seasonal (b) Occasional

(c) Regular (d) Endemic

180. The pest which occure during perticular part of year this pest is called aspest.

(a) Occasional (b) Regular

(c) Endemic (d) Seasonal pest

181. The pest which is normally causing a loss ranging 5 to 10 per cent are said to be__________pest.

(a) Minor (b) Major

(c) Negligible (d) Obligate

182. The Integrated control term first time used by__________scientist.

(a) Bacher and Bacon (b) Smith

(c) Hackel (d) Stern

183. The Integrated control was first defined by__________scientist in 1959.

(a) Steinhaus (b) Stern *et al.*

(c) Hackel (d) Bacher

184. The lowest pest population density that will cause economic damage is called as__________.

(a) GEP (b) EIL

(c) ETL (d) RL

Answer Key

1–b	23–a	45–b	67–a	89–a	111–a
2–d	24–b	46–a	68–a	90–a	112–c
3–c	25–a	47–c	69–b	91–a	113–a
4–a	26–c	48–b	70–c	92–b	114–b
5–d	27–a	49–d	71–b	93–d	115–a
6–c	28–a	50–a	72–c	94–c	116–c
7–b	29–d	51–b	73–c	95–c	117–b
8–c	30–c	52–c	74–c	96–b	118–a
9–b	31–a	53–b	75–c	97–a	119–a
10–d	32–b	54–b	76–b	98–b	120–b
11–b	33–d	55–a	77–a	99–d	121–c
12–d	34–b	56–b	78–a	100–b	122–d
13–b	35–c	57–c	79–c	101–a	123–a
14–b	36–d	58–c	80–a	102–b	124–c
15–b	37–c	59-a	81–d	103–c	125–c
16–c	38–a	60–c	82–d	104–d	126–a
17–d	39–b	61–d	83–b	105–d	127–b
18–b	40–a	62–a	84–d	106–c	128–c
19–c	41–a	63–c	85–d	107–a	129–b
20–b	42–d	64–a	86–c	108–a	130–c
21–b	43–b	65–b	87–d	109–d	131–d
22-d	44–a	66–c	88–a	110–c	132–b

133–c	**142–b**	**151–a**	**160–b**	**169–b**	**178–b**
134–c	**143–c**	**152–a**	**161–b**	**170–c**	**179–c**
135–c	**144–c**	**153–a**	**162–c**	**171–b**	**180–d**
136–c	**145–c**	**154–b**	**163–a**	**172–a**	**181–a**
137–c	**146–c**	**155–b**	**164–b**	**173–d**	**182–a**
138–b	**147–c**	**156–b**	**165–b**	**174–c**	**183–b**
139 –b	**148–d**	**157–d**	**166–d**	**175–c**	**184–b**
140–c	**149–a**	**158–c**	**167–d**	**176–b**	
141–b	**150–c**	**159–d**	**168–b**	**177–d**	

Chapter 9
Biological Control

1. *Aenasius bambawalei* Hayat: Hymenoptera: Chalcidoidea: Encyrtidae
2. *Aenasius bambawalei endoparasitoid belonging to encrytidae and parasitising Phenacoccus solenopsis.* (IARI Ph. D. 2011)
3. Commonwealth Institute of Biological Control is headquartered in: Trinidad
4. Cteptoparasitism is a type of parasitism in which the adult parasitoids preferentially appropriate for its own progeny of the previously parasitized host of another parasitoid.
5. *Eriborus trochanteratus* is an exotic larval parasite on coconut black headed caterpillar.

Multiple Choice Questions

1. When many species of parasite attack on single individual of host it is known as__________.

(a) Superparasitism (b) Multiparasitism

(c) Hyperparasitism (d) Simple parasitism

2. When number of individual of same species of parasite attack on single host is known as__________.

(a) Multiparasitism (b) Superparasitism

(c) Hyperparasitism (d) Simple parasitism

3. The parasite that attack on another parasite is called as __________.

(a) Primary parasite (b) Secondary parasite

(c) Tertiary parasite (d) Hyperparasite

4. ____________stage of parasitoid attack on host.

(a) Egg (b) Larval

(c) Adult (d) All of these

5. Which one of the following is an egg parasitoid?

(a) *Trichospilus pupivora* (b) *Goniozus nephantidis*

(c) *Bracon bravicornis* (d) *Trichogramma* spp

6. ____________ is laboratory host for multiplication of egg parasitoids.

(a) *Corcyra cephalonica* (b) DBM

(c) *Helicoverpa armigera* (d) All of these

7. *Opuntia inermis* (Prickly pear) is controlled with____________insect.

(a) *Dactylopius cocus* (b) *Neochetina bruchi*

(c) *Cactoblastic cactarum* (Moth borer) (d) *Zygogramma bicolorata*

8. Cactus, *Opuntia dillenii* is controlled by____________insect.

(a) *Zygogramma bicolorata* (b) *Neochetina bruchi*

(c) *Dactylopius tomentosus* (d) *Cactoblastic cactarum*

9. Water hyacinth is effectively controlled by____________insect.

(a) *Zygogramma bicolorata* (b) *Dactylopius tomentosus*

(c) *Neochetina bruchi* (d) All of these

10. Agostino bassi worked on____________disease of silk worm.

(a) Yellow muscardine (b) Green muscardine

(c) White muscardine (d) Red muscardine

11. The term microbial control was firstly given by ____________.

(a) Smith, 1959 (b) E. A. Stainhaus, 1949

(c) Stebbing, 1949 (d) Clerk, 1941

12. *Bt* was first isolated by ____________in 1902 from larvae of silk worm.

(a) Smith (b) Ishiwata

(c) Agostino bassi (d) Pradhan

13. Which of the following toxin produced by Bt?

(a) Delta-edotoxin (b) Destruxins

(c) Beauvericin (d) All of these

14. *Bacillus popilliae* bacteria first isolated from larvae of ____________.

(a) White grub (b) Meditarian flour moth

(c) Japanese beetle (d) All of these

15. __________family of virus is most important to control insect pest.

(a) Astrovoridae (b) Bornaviridae

(c) Arterividae (d) Baculoviridae

16. __________virus is used to control *Chilo infuscatellus*.

(a) SlNPV (b) CiGV

(c) HaNPV (d) TnjGV

17. One larval equivalent have __________POBs/ml.

(a) 6×10^{9} (b) 3×10^{9}

(c) 6×10^{11} (d) 3×10^{11}

18. __________is first viral insecticide product of HaNPV.

(a) Doom (b) Dipel

(c) Elcar (d) Biolep

19. *Steinernema* and *Heterorhabditis spp.* of nematode are used to control __________insect pest.

(a) Whitefly (b) Soil dwelling

(c) Scale insect (d) Bugs

20. Genetically modified brinjal is developed for control of__________.

(a) Shoot and Fruit borer, *Leucinodes orbonalis* (b) Bollworm

(c) Spodoptera (d) Whitefly

21. __________is parasitic nematodes used for control of insects.

(a) *Globodera* (b) *Meloidogyane*

(c) *Steinernema feltiae* (d) *Nacobbus* sp.

22. The term biological control first used by__________scientist in 1919.

(a) Smith (b) Bacon

(c) Berliner (d) Stern

23. The__________parasite notice laying eggs in the pupae of cabbage butterfly.

(a) *Trichograma chilonis* (b) *Apentalis glomeratus*

(c) *Trichograma minutum* (d) *Bracon brevicornis*

24. The__________ parasitoid used to control Sugarcane pyrilla.

(a) *Trichogramma chilonis* (b) *Apentalis glomeratus*

(c) *Epiricania melanoleuca* (d) *Trichograma minutum*

25. __________bird first used as bio agent in 1762.

(a) Parrot (b) Crow

(c) Indian myna (d) Pigeon

26. The___________parasitoid used to control potato tuber moth.

(a) *Epricania melanoleuca* (b) *Trichogramma chilonis*

(c) *Chelonus blackburni* (d) *Apentalis glomeratus*

27. ___________bird is used as natural enemy to control red locust.

(a) Owl (b) Crow

(c) Sparrow (d) Indian myna

28. ___________nematode used for control of Japanese beetle.

(a) *Neoplectana bombasis* (b) *Nepolectana carpocapsae*

(c) *Neoplectana glaseri* (d) *Heterorahabitis*

29. ___________bacteria having symbiotic relationship with nematode *Neoplectana carpocapsae.*

(a) *Bacillus nematophilus* (b) *Achromobacter nematophilus*

(c) *Bacillus popillae* (d) *Bacillus thuringiensis*

30. Bt is isolated from___________pest.

(a) Hawk moth (b) Mediterranean flour moth

(c) Rice moth (d) Black moth

31. ___________toxins is produced by *Bascillus thuringiensis*.

(a) Alpha toxin (b) Beta toxin

(c) Delta endotoxin (d) Gamma toxin

32. ___________fungus used against grubs of rhinoceros beetle.

(a) *Apentellus flaves* (b) *Metarrhizium anesopliae*

(c) *Verticellium* (d) *B. bassiana*

33. ___________fungus used to control sucking pest.

(a) *Apaenteleus flavipes* (b) *Lecanicillium lecanii*

(c) *Aspergillus flavus* (d) *N. rileyi*

34. ___________virus is used to control the *Helicoverpa armigera*.

(a) GV (b) CPV

(c) TPV (d) NPV

35. To control the *Lantana camera,* which one of the fly is used?

(a) Stemfly (b) Fruitfly

(c) Seedfly (d) Blackfly

36. The parasitoid attacks on__________stage of insect host.

(a) Egg (b) Larval

(c) Adult (d) All of these

37. Predator is host specific to one species of prey is called as__________predator.

(a) Facultative (b) Afacultative

(c) Sedentary (d) Obligate

38. *Rhodolia cardinalis* used for control of__________pest.

(a) *Aphis gossypii* (b) *Icerya purchessi*

(c) *Earis vitella* (d) *Aphis dorsata*

39. __________microorganisms are host specific.

(a) Fungi (b) Virus

(c) Bacteria (d) Protozoa

40. __________is example of external parasites.

(a) Lizard and Flies (b) Mites and Flies

(c) Frog and Flies (d) Ticks and Flies

Answer Key

1–b	8–c	15–d	22–a	29–b	36–d
2–b	9–a	16–b	23–b	30–b	37–d
3–d	10–c	17–a	24–c	31–c	38–b
4–b	11–b	18–c	25–c	32–b	39–b
5–d	12–b	19–b	26–c	33–b	40–d
6–a	13–a	20–a	27–d	34–d	
7–c	14–c	21–c	28–c	35–c	

Match the Pairs

(A)

Fungus		Pest
1. *Metarrhizium anisopliae*	a.	Grasshopper
2. *Entomophthora grylli*	b.	Rhinoceros beetle
3. *Aspergillus flavus*	c.	Pyrilla, Rice Skipper
4. *Metarrhizium flavoridae*	d.	Caster Semilooper, Japanese beetle
5. *Beauveria bassiana*	e.	Locust, Grasshopper

(B)

Fungus		Pest
1. *Verticillium lecanii*	a.	Aphids, Jassids, Hoppers, Scale insect
2. *Entomophthora lecanii*	b.	Coffee green scale, *Coccus viridis*
3. *Nomuraea rileyi*	c.	*Spodoptera exigua, S. litura, H. armigera*
4. *Entomophthora sphaerosperma*	d.	Cabbage Butterfly
5. *Neozygites fresenii*	e.	*Aphis gossypii*

(C)

Viruses		Pest
1. HaNPV	a.	*Chilo infuscatellus*
2. SlNPV	*b.*	*Helicoverpa armigera*
3. CiGV	c.	*Spodoptera litura*
4. CpNPV	d.	*Oryctes rhinoceros*
5. OrNPV	e.	*Cydia pomonella*

(D)

Bacteria		Pest
1. Bt. var kurstaki	a.	Mosquito larvae
2. Bt.var israelensis	b.	Lepidopteran larvae
3. *Bacillus papillae*	c.	Coleoptera
4. Bt.var tenebrionis	d.	White grubs
5. *Bacillus sphaericus*	e.	Culex Mosquito

(E)

Protozoa		Insect
1. *Nosema apis*	a.	Grasshopper
2. *Malpighamoeba mellificae*	b.	Honey bee
3. *Nosema bombycis*	c.	Silkworm
4. *Farinocystis triboli*	d.	*Tribolium castaneum*
5. *Nosema locustae*	e.	Honey bee

(F)

Protozoa	Insect
1. *Romanomermis culicivorax*	a. Mosquito
2. *Steinernema feltiae*	b. Codling moth
3. *Vairimorpha necatrix*	c. *Spodoptera litura, Helicoverpa armigera*
4. Flagellate protozoa	d. Termite
5. *Perezia pyraustae*	e. European corn borer

Answer Key

A	**1- b**	**2- a**	**3- c**	**4- e**	**5- d**
B	**1- a**	**2- b**	**3- c**	**4- d**	**5- e**
C	**1- b**	**2- c**	**3- a**	**4- e**	**5- d**
D	**1- b**	**2- a**	**3- d**	**4- c**	**5- e**
E	**1- e**	**2- b**	**3- c**	**4- d**	**5- a**
F	**1- a**	**2- b**	**3- c**	**4- d**	**5- e**

Chapter 10
Toxicology

1. Dark yellow colour label on container of pesticide: highly toxic (ICAR-SRF 2011).
2. Endosulfan the ratio of alpha and beta isomers: 70:30.
3. First time quarantine started in India at Mumbai.
4. Causes of OP poisoning: Anorexia, Nausea, Bradycardia.
5. Sevidol is mixture of sevin and lindane.
6. Triton X 100 used as emulsifier.
7. A safety factor ideally 100 folds is incorporated while fixing MRL for each commodity.
8. Apolar insecticides penetrate cuticle much faster than polar insecticides.
9. Insecticides with high vertebrate selectivity ratio are more toxic to insect than vertebrate.
10. Flood jet nozzle: This type throw out the spray in the form of a jet of coarser droplet. Since the jet falls at right angle to the surface, drifting of the chemical is minimised. This is used for herbicide spray.
11. Atropine sulphate (2-4 mg intramuscular/intravenous administration) antidote for: Carbamates.
12. Enobarbital, Diphenylhydration antidote for: Synthetic pyrethoid.
13. Atropine sulphate (2-4 mg intramuscular/intravenous administration) or PAM (Pyridine-Z aldoxime-N-methliodide) antidote for: Organophosphate Compounds.

14. Phytoecdysone: A plant product that mimics ecdysone.
15. Insectidal activity of isomers of DDT falls as: para-para > meta para > meta orthro > orthro orthro.
16. First systemic inorganic insecticide: Sodium selenate.
17. Insecticide used as stem injection for management of coconut black headed caterpillar: monocrotophos.
18. Universal antidote: 7 parts of activated charcoal + 3.5 part of MgO_6 + 3.5 part Tannic acid.
19. The insecticide formulated as 100 per cent EC: DDVP.
20. DDT is oxidized to Dichloro diphenyl acetic acid (DDA) in Rats.
21. DDT is oxidized to Dicofol in Drosophila.
22. DDT is oxidized to Dichloro benzophenone (DBP) in cockroaches.
23. DDT is oxidized to Dichlorodiphenyl dichloroethylene (DDD).
24. A contact insecticide should have the low partition coefficient.
25. The insecticidal properties of BHC were discovered by Dupit, Raucourt and Slade (1942).
26. In housefly the resistance to DDT is due to recessive gene called kdr on chromosome 3.
27. The state which ranks first in consumption of pesticides: Andhra Pradesh.
28. Juvenile hormone: third generation of pesticide (ICAR-NET 2013).
29. Anti juvenile hormone: fourth generation pesticide (ICAR-NET 2013).
30. The sprayer suitable for use in plantations: Rocker sprayer (ICAR-NET 2013).
31. The sprayer suitable for use in fruit crops: Foot pump sprayer (ICAR-NET 2013).
32. Aluminium phosphide available under various brand names, *e.g.* Celphos, Fumitoxin, Phostoxin, Talunex and Quick Phos.
33. Aluminium phosphide reacts with water or acids to release phosphine gas.

    $$AlP + 3\,H_2O \rightarrow Al\,(OH)_3 + PH_3$$

    $$AlP + 3\,H^+ \rightarrow Al^{3+} + PH_3$$

34. Melting point of Aluminium phosphide is 2530 °C.

35. Aluminium phosphide also known as Aluminium (III) phosphide and Aluminium monophosphide.
36. The dust hopper should be filled with dust to its $3/4^{th}$ capacity.
37. Recently banned insecticide is Malathion.
38. The insecticide Fenvolerate belongs to synthetic pyrethroid group.
39. Vector (RH-787) is used as rodenticide.
40. The Central Insecticide Laboratory (CIL) is located at Faridabad.
41. The trade name of phosphamidon is Dimecron.
42. DDT was synthesized by Zeidler, 1874 (IARI Ph. D. 2011).
43. The movement of insecticide away from the soil surface along with water vapour: Wicks effect.
44. Chemicals which can induce congenital malformation: teratogenic.
45. Leaving the pest below economic threshould level for survival of natural enemies during insecticidal application: Pest residue.
46. The boiling point of methyl bromide: 20°C.
47. Otruorixtfeb (IGR) is effective against: cell sap sucker.
48. First chitin synthasis inhibitor developed as insecticide: Diflubenzuron.
49. "Match" is trade name of Lufenuron (Chitin synthesis inhibitor).
50. Lufenuron: chitin synthesis inhibitor (IARI Ph. D. 2011).
51. "Applaud" is trade name of Buprofenzin.
52. Buprofezin is chitin synthesis inhibitor.
53. "Nomalt" is trade name of Teflubenzuron.
54. Benzyl benzoate is used as repellent against mite and ticks.
55. RD -9: neem based formulation (100 per cent EC Azadirachtin).
56. Indian beech is common name of *Pongamia pinnata*.
57. Active ingredients of Palas (*Butea monosperma*): chalcone, aurones (termiticidal properties).
58. Hashimoto and Okaichi proposed the chemical stracture of neristoxin (1962).
59. Institute of Pesticide Formulation Technology is located at Gurgaon, Haryana.

60. Imidacloprid: 1st commercial insecticide of Chloronicotinyl group.
61. Fipronil is popular with the brand name Regent ®.
62. Insecticide "Pymetrozine": blocks stylet penetration of sucking insects.
63. Indoxacarb is available in market with trade name as Auvant ® and Auvant EC.
64. Formamidines possess ovicidal activity against lepidopteran pest.
65. Formamidines inhibits the enzyme mono amine oxidase resulting in the accumulation of biogenic amines *i.e.* octopomine.
66. Carbon sulphide controls: nematode.
67. Activation hormone(AH) is secreted by neuro secretory cells (NSC).
68. House fly is mainly used for testing of chemosterilants.
69. Discriminating dose: A dose that is used to discriminate between resistant and susceptible individuals and is not directly related to field efficacy.
70. Mode of Action (MOA): The biochemical process by which a pesticide disrupts normal pest biology usually resulting in the death of the pest. Normally this is a target binding site or a key biological process.
71. Multisite compound: a compound that affects more than one target site.
72. Pro-insecticide: it is metabolized into an active insecticide after entering the host.
73. Pro-insecticide: cartap, padan, thiocyclam, Chlorfenapyr.
74. Chlorfenapyr: halogenated pyrrole compound.
75. Chlorfenapyr: disrupting the production of Adenosine triphosphate (resulting in disruption of production of ATP).
76. Formamidines: posses ovicidal activity against lepidopteran pest.
77. Chess®, Reley® : commercial product of pymetrozine (USA market).
78. International Institute of Biotechnology and Toxicology (IIBAT): Padappai (Tamil Nadu).
79. LD_{50}: The amount of insecticide per unit weight which will kill 50 per cent of the particular organism or insect.
80. LC_{50}: The concentration of insecticide required to kill 50 per cent of the given organism or insect.
81. LC_{50} is expressed in PPM (1/1,000,000) or Percentage (1/100).

82. ED_{50} and EC_{50} are the dose or concentration of the chemical (IGR) required to affect 50 per cent of population and produce desired symptoms in them.

83. Only 1 per cent of the pesticide applied to crop reaches the target. The remaining 99 per cent contaminate soil, water, air, food, forage, etc.

84. Chemosterilants: A chemical compound which deprive off insect from their ability to reproduce the progeny.

85. Aluminium phosphide: fumigant (ICAR-NET 2013).

86. Methyl bromide: ozone depleting compound (ICAR-NET 2013).

87. Applaud: active ingredient -Buprofenzin (ICAR-NET 2013).

88. Summer oil: These are highly refined and possess a U.R. (unsulfonatable residue) rating 90-96 per cent.

89. Dormant oil: These are heavy and less refind and possess a U. R. rating 50-90 per cent.

90. Nicotine: Nicotine- 1, 3-(I-Methylpyrrolidyl) pyridine- $C_{10}H_{14}N_{12.}$

91. Nicotine sulphate is banned for use in India.

92. Characteristic action of pyrethroid: rapid paralysis (knock down).

93. The rotenone was first isolated in 1895 by Geoffrey.

94. Tephrosin: toxic principles are found in leaves and seed of *Tephrosia vogelii*, in roots of *T. toxicaria*, and in stem of *T. macropoda.*

95. *Lobelia excels*: plant belonging to companulaceae, has been effective against aphid on snake gourd and cowpea, tinged on brinjal and mites on castor and bhendi.

96. Baker and Maughum described first acaricidal properties of Dicofol in 1956.

97. Excitatory neurotransmitters: They result in the signal being sent on through the synapse to a connecting neuron.

98. Inhibitory neurotransmitters: They result in the reaction being blocked from travelling through the synapse to a connecting neuron.

99. GABA: inhibitory neurotransmitter.

100. The mosquito coils consist of a base made of wood powder, starch, coconut shell powder, dyestuff, glues and burning regulators.

101. In case of mosquito mat vaporisers consist of a base made of cellulose fibre board or cardboard or plastic mat. (IARI Ph. D. 2011).

102. The mosquito mat vaporisers consist an active ingredient and trace of perfumes, dyestuffs and evaporation inhibitors, hydrocarbon solvents.

103. The only product registered in India for mosquito control in the form of agarbatti: pyrethrum 0.3 per cent.

104. In most of the mosquito mat vaporiser systems the temperature on the heater surface is between140°C to 190°C.

105. The operation temperature for liquid vaporizer: 125°C to 140°C.

106. Bait granules are used for cockroach and fly control.

107. Gel baits: latest technological development in cockroach control.

108. Aerosols cans are designed to release small droplet suspensions into air to control flying insects such as mosquitos and flies: flying insect killer (FIK).

109. Multi Insect killer (MIK) aerosols: aerosols developed to control both flying insects and crawling insects.

110. MIK aerosols contain a knock down insecticide and a residual insecticide in combination.

111. Fipronil belongs to group phenyl pyrazoles. (ICAR-SRF, 2013).

112. Aerial spray of insecticide relates a big issue (cashew crop): Endosulfan.

113. Oxadiazine insectide: Indoxacarb (IARI Ph. D. 2011).

114. Triple action nozzle: (Hollow cone + solid cone).

115. The amount of initially laid down pesticide on the surface is called deposit.

116. The portion of pesticide left from initial deposit is called surface or effective residues.

117. Breakdown product of the pesticides, which are stable and create as many problems as the original compound, are called as terminal residue.

118. The physical and chemical changes that occur for a pesticide within an organism: metabolism.

119. Dissipation follows the low of "first-order kinetics", whereas persistence does not follow the same low.

120. DT_{50}: Dissipation time 50 per cent time required for one-half of the initial quantity or concentration of a pesticide to dissipate form a system.

121. Half-life ($t_{0.5}$): time taken for the concentration of a pesticide in a compartment to decline one half.

122. Liquid fumigants for storage: Ethylene dibromide (EDB), Methyl bromide (MB).

123. Storicide II- IGR used as grain protectant.

124. Phenobarbitol: antidote for organophosphate.

125. Similar joint action of poisons: quantal responses to mixtures of poisons acting similarly could be predicted from the separate toxicities of the components of the mixtures.

126. *kdr* (knock-down resistance): interferes with the sodium channel in nerve cells, (common mechanism used for resistance to DDT and pyrethroids), *e.g.* in *Anopheles gambiae, Blattella germanica.*

127. *MACE* (modified acetylcholinesterase): modifies the structure of acetylcholinesterase so that it is no longer affected by the insecticide.

128. *Rdl* (resistance to dieldrin) is a point mutation that reduces dieldrin binding at the GABA receptor (dieldrin resistance in *Anopheles quadrimaculatus* mosquitoes).

129. FRAC: Fungicide Resistance Action Committee.

130. FRAG: Fungicide Resistance Action Group.

131. HRAC: Herbicide Resistance Action Committee.

132. IRAC: Insecticide Resistance Action Committee.

133. IRAG: Insecticide Resistance Action Group.

134. RRAC: Rodenticide Resistance Action Committee.

135. Resistance mechanism: biological processes used by the pest to avoid the lethal action of the pesticide. Resistant organisms may have more than one resistance mechanism.

136. Emamectine benzoate derived from *Streptomyces avermitilis*. (IARI Ph. D. 2011).

137. Emamectine benzoate effective against lepidopteran pests.

138. 'K eating insect powder': dried powder of pyrethrum flower.

139. The first synthetic pyrethriod 'allethrin' was made by Schechter et al.

140. Cypermethrin: mixture of eight isomers.

141. First soil active pyrethroid: tefluthrin.

142. RH-5849: first compound to bind with ecdysteroid receptor – non steroidal hyperecdysonism syndrome.

143. A new insecticide (chemical class- spirocyclic phenyl substituted tetronic acid) effective aginst white fly and spider mite: Spiromeslfen (IARI Ph. D. 2011).

144. Avermectines are chloride channel activators. (IARI Ph. D. 2011).

145. Metaldehyde commonly used for management of snail. (ICAR-NET 2013).

146. JH is also known as ectohormone.

147. Hollow cone nozzle produces spary in the form of hollow cone with ring shaped impact area.

148. Pest resurgences mostly noticed in Homoptera (about 44 per cent).

149. (AC 24055) has been most widely used triazene which is odourless, tasteless, non toxic chemical which inhibit feeding in chewing insect.

150. Baygon: systemic antifeedent.

Multiple Choice Questions

1. **Chemical which kill the mites is called as________.**
   (a) Molluscicide (b) Mitigocide
   (c) Acaricide (d) Insecticide

2. **Chemical which kill the molluscs is called as________.**
   (a) Rodenticide (b) Mitigocide
   (c) Molluscicide (d) Insecticide

3. **Pesiticides which are derivatives of naturally occurring biochemicals, bacteria, fungi, viruses etc. are called as________Pesticide.**
   (a) Biorational (b) Pheromone
   (c) Chemical (d) Synthetic

4. **First use of________insecticide is done for control of Colorado potato beetle in 1867.**
   (a) Paris green (b) DDT
   (c) BHC (d) TEPP

5. **In________insect, insecticide resistance was firstly observed in India.**
   (a) Singhara beetle (b) DBM
   (c) *Helicoverpa armigera* (d) *Aphis gossypii*

6. **________formulation is suitable for soil application of insecticide.**
   (a) Dust (b) SL
   (c) Granular (d) WSP

7. **Dust have________ particle size.**
   (a) >100 micron (b) <100 micron
   (c) <50 micron (d) >50 micron

8. ________formulations of insecticide require to be apply in the early morning.

(a) Liquid (b) Dust

(c) Granular (d) WP

9. Carriars used in dust formulation are________.

(a) Attapulgite (b) Kaolins

(c) Sulphur (d) All of these

10. Granular formulation of insecticide prepared by________method.

(a) Agglomeration technique (b) Spray impregnation

(c) Stickon technique (d) All of these

11. Microgranular insecticides have partial size________micron.

(a) <100 (b) <300

(c) <500 (d) 700

12. Granular formulations have________ per cent toxicants.

(a) 2 to10 (b) 10-15

(c) <5 (d) >15

13. ________compounds are used as emulsifying agent for preparing EC formulations.

(a) Soaps (b) Organic amines

(c) Both (a) and (b) (d) Lime

14. Store grain pests are mostly controlled by________method of pest control.

(a) Fumigation (b) Aerosols

(c) Fog (d) All

15. The phenomenon in which toxicity of mixture of insecticide is greater than that of sum of individual insecticide is known as________.

(a) Synergism (b) Antagonism

(c) Similar action (d) Joint action

16. Adjuvents that are used to stick the pesticide to foliage that prevent run off of spray fluid are known as________.

(a) Surfactant (b) Stickers

(c) Wetting agent (d) Spreaders

17. Which of the following insecticide are cholinesterase inhibitors?

(a) O. P. (b) Carbamates

(c) Arsenicals (d) Both (a) and (b)

18. ________group of insecticide act as mimics of acetylcholine.

(a) O.P (b) O.C

(c) Neonicotinoids (d) Pyrethroids.

19. Which of the following insecticide are chlorine channel activators?

(a) Avermectins (b) O.C

(c) Carbamates (d) Both (a) and (b)

20. ________group of insecticides inhibit the activity of GABA.

(a) O.C (b) O.P

(c) Carbamate (d) All of the above

21. Which of the following is first photo stable pyrethroid?

(a) Alletrin (b) Cypermethrin

(c) Permethrin (d) Deltamethrin

22. ________insecticides have quick knock-down effect.

(a) O.P (b) O.C

(c) Pyrethroids (d) None of these

23. Which of the following insecticides act on sodium channels?

(a) Pyrethroids (b) O.C

(c) O.P (d) Carbamates

24. Heavy oils and tar oils can kill insect by________.

(a) Asphyxiation (b) Protoplasmic disintegration

(c) Dehydration (d) None of these

25. ________toxicant act as protoplasmic poison.

(a) Heavy metals (b) O.P

(c) O.C (d) Carbamates

26. Sulphur act as________.

(a) Molluscicide (b) Acaricide

(c) Rodenticide (d) All of these

27. ________are example of animal origin insecticide.

(a) Cartap hydrochloride (b) Nereistoxin

(c) Neonicotinoide (d) Both (a) and (b)

28. Nereistoxin isolated from________.

(a) Tobacco (b) Marine annelids

(c) Fish (d) All of these

29. Chemical formula of Nicotine is_________.

(a) $C_{10}H_{14}N_2$ (b) $C_3H_6O_3$

(c) $C_{14}H_{10}N_2$ (d) $C_6H_{12}O_6$

30. Insecticidal property in Custard apple due to_________.

(a) β-pinene (b) Gammaexne-D

(c) α-pinene (d) All of these

31. Nicotine content in the leaves of *Nicotina tobaccum* is_________ per cent.

(a) 2 to 5 (b) 20

(c) 30 (d) <2

32. Which of the following alkaloids present in tobacco?

(a) Anabasine (b) Nornicotine

(c) Nicotine (d) All of these

33. _________synergist are used in pyrethroids.

(a) Piperonyl butoxide (b) Sesamin

(c) Creosote oil (d) Both (a) and (b)

34. Pyrethroid insecticides are_________poison.

(a) Contact (b) Stomach

(c) Nerve (d) Both (a) and (c)

35. Rotenone is isolated from roots of_________.

(a) *Derris elliptica* (b) *Lonchocarpus utilis*

(c) *Lonchocarpus uruca* (d) All of these

36. *Lonchocarpus* spp. have_________ per cent rotenone in its roots.

(a) 8-11 (b) 1-3

(c) 3-5 (d) 4-7

37. _________alkaloid present in stem and roots of Ryania.

(a) Nimbecidine (b) Ryanodine

(c) Tephrosin (d) All of these

38. Control of store grain pest in storage achieved by use of_________.

(a) *Acorus calamus* (b) *Annona Squamosa*

(c) *Pongamia glabra* (d) All of these

39. DDT was synthesized by_________in 1874.

(a) Paul Muller (b) A. P. W.Dupire

(c) O. Zeidlar (d) All of these

40. Insecticidal properties of DDT discovered by_________in 1939.

(a) O. Zeidlar (b) A. P. W. Dupire

(c) Paul Muller (d) Michael Faraday

41. _________of following is effective acaricide.

(a) Dicofol (b) HCH

(c) DDT (d) Endosulfan

42. _________form of DDT is non toxic.

(a) DDE (b) Dicofol

(c) DDD (d) DDA

43. DDT is converted or oxidized in to _________ form in insect.

(a) Dicofol (b) DDE

(c) DDD (d) DDA

44. _________insecticide has very high accumulation rate in fatty tissue.

(a) DDT (b) Endosulphan

(c) DDE (d) Phosphamidon

45. Which of the following are used as synergist for pyrethrin insecticide?

(a) Sulfoxide (b) Sesamex

(c) Pipronyl butoxide (d) All of these

46. HCH was first synthesized by _________in 1825.

(a) Michel Farady (b) Paul Muller

(c) Othmer Zeidler (d) W. Finkenbrink

47. _________had given the insecticidal properties of HCH.

(a) Michel Farady (b) Vander Linden

(c) A. P. W. Dupire (d) Smith C. F.

48. Who isolated the gamma isomer of HCH?

(a) Paul Muller (b) Othmer Zeidler

(c) Vander Linden (d) W. Finkenbrink

49. 99 per cent pure gamma isomer of HCH is called as _________.

(a) Chlorodane (b) Linden

(c) BHC (d) Endosulphan

50. Which of the following are cyclodine insecticides _________.

(a) Aldrin (b) Dieldrin

(c) Endosulphan (d) All of these

51. Insecticidal properties of endosulphan given by ________ in year 1956.

(a) W. Finkenbrink (b) Michel Farady
(c) Paul Muller (d) Orthmar Zeidler

52. ________ insecticide have ovicidal action.

(a) Ethion (b) Profenophos
(c) Both (a) and (b) (d) Aldrien

53. ________ are non systemic insecticides.

(a) Malathion (b) Endosulphan
(c) Diazinon (d) All of these

54. ________ insecticide is safe for controlling pest of animals.

(a) Malathion (b) Phorate
(c) DDT (d) Endosulphan

55. Which of the following are acaricides?

(a) Dicofol (b) Sulphur
(c) Milabemectin (d) All of these

56. ________ insecticide is safest to predator and parasitoides.

(a) Phosalone (b) Phorate
(c) DDT (d) Endosulphan

57. Insecticidal properties of methyl parathion given by ________.

(a) W. Finkenbrink (b) Chamberlain
(c) Paul Muller (d) G. Schrader

58. Trichlorfan is dehydrochlorinated in to ________ in insect body.

(a) Dichlorovas (b) Endosulphan sulphate
(c) DDE (d) Ethaphon

59. Insecticidal properties of carbaryl given by ________.

(a) Paul Muller (b) Othmer Zeidler
(c) H. L. Haynes (d) W. Finkenbrink

60. ________ insecticide is effectively used to control soil inhibiting insect and nematodes.

(a) Endosulphan (b) Carbofuran
(c) Chloropyriphos (d) None of these

61. ________ is an oxime carbamate act as molluscicide.

(a) Thiodicarb (b) Methomyl
(c) Both (a) and (b) (d) Chloropyriphos

62. **Pyrethroids are_________in action.**
   (a) Contact (b) Systemic
   (c) Both (a) and (b) (d) Fumigant

63. **_________is an acute rodenticide.**
   (a) Zinc phosphide (b) Warfarin
   (c) Both (a) and (b) (d) Dichlorovos/DDVP

64. **_________is a chronic rodenticide.**
   (a) Warfarin (b) Zinc phosphide
   (c) Bromadiolone (d) None of these

65. **Which of following are chemosterilant?**
   (a) Tretamine (b) Apholate
   (c) TEPA (d) All of these

66. **_________is used in mosquito control coils.**
   (a) Endosulphan (b) Phosphamidon
   (c) Delta allethrin (d) DDT

67. **Fipronil belongs to _________group of insecticide.**
   (a) Oxadiazine (b) Phenyl pyrazole
   (c) Pyrazole (d) Neonicotinoids

68. **_________is rodenticide with no bait shyness and anticoagulant type.**
   (a) Bromodiolone (b) Warfarin
   (c) Zink phosphate (d) All of these

69. **_________ is most popular molluscide**
   (a) Metaldehyde (b) Bromadiolone
   (c) Aluminum phosphide (d) Fenvalerate

70. **_________are chemical used to cure the cases of insecticide toxicity.**
   (a) Antidotes (b) Suppliments
   (c) Adjuvants (d) Absorbents

71. **Universal antidotes have _________constituents.**
   (a) Tannic acid (b) Magnesium oxide
   (c) Activated charcoal (d) All of these

72. **_________antidote is given to cure organochlorine poisoning.**
   (a) Phenobarbital (b) Atropine
   (c) $KMnSO_4$ (d) Magnesium sulphate

73. ________antidote is used to cure organophosphate insecticide poisoning.

(a) Phenobarbital (b) Atropine

(c) Morphine (d) Both (a) and (b)

74. Antidote used for Bromadiolone is ________.

(a) Vitamin K (b) Atropine

(c) Phenobarbital (d) Morphine

75. Cloth moth can be controlled by ________.

(a) Para dichlorobenzene (b) Napthalene

(c) Malathion (d) Both (a) and (b)

76. Neonicotinoids insecticide includes ________insecticides.

(a) Acetamiprid (b) Imidacloprid

(c) Clothianidin (d) All of these

77. Which one of the following insecticide is fumigant in action?

(a) Lambda cyhalothrine (b) Deltamethrin

(c) DDVP (d) Endosulphan

78. Controlled droplet application (CDA) technique used in ________application of insecticide.

(a) ULV (b) UV

(c) HV (d) Fumigants

79. If higher the LD_{50} value ________is toxicity of insecticides.

(a) Lower (b) Higher

(c) Neutral (d) None of these

80. Rotary dusters can cover________hacter per day.

(a) 1-1.5 (b) 2-2.5

(c) 3-3.5 (d) 4-4.5

81. Spraying is effective when number of droplets per cm^2 is ________.

(a) 20 (b) 30

(c) 50 (d) 80

82. ________part of spray pump convert spray fluid in to droplet.

(a) Pump (b) Cut of device

(c) Nozzle (d) Boom

83. ________type of nozzle used to spraying insecticide.

(a) Solid cone (b) Hallow cone

(c) Flat fan (d) Jet

84. **________type of nozzle is used to spray a weedicide.**
   (a) Hallow cone (b) Flat fan
   (c) Solild cone (d) Jet

85. **Pneumatic knapsack sprayer has working pressure ________kg/cm^2.**
   (a) 2-4 (b) 4-7
   (c) 10 (d) 14-18

86. **ULV applications have droplet size ________micron.**
   (a) 40-150 (b) 20-150
   (c) 50-150 (d) 60-150

87. **________area per day covered with ULV applicator.**
   (a) 8ha (b) 12ha
   (c) 15ha (d) 17 ha

88. **________chemical used in bird scarer.**
   (a) Calcium cynide (b) Calcium carbonate
   (c) Calcium carbide (d) Calcium nitrate

89. **________equipment is used as mechanical device to control locust.**
   (a) Fogger (b) Bird scarer
   (c) Flame thrower (d) Soil injector

90. **________is the part on which number of nozzles are attached.**
   (a) Lance (b) Presurre pipe
   (c) Hose pipe (d) Boom

91. **Insecticidal chemicals in pyrethrums are________.**
   (a) Pyrethrin-I and II (b) Jasmoline-I and II
   (c) Cinerin-I and II (d) All of these

92. **________is first synthetic pyrethroid synthesized in 1949.**
   (a) Allethrin (b) Cypermethrin
   (c) Allethroid (d) decamethrin

93. **________artificialy synthesized Azadirectin in 2007.**
   (a) Dr. Steven V.Leq (b) Shuk A.
   (c) Pradhan R. (d) Joshi A. S.

94. **Avermectins are derived from soil actinomycete________.**
   (a) *Streptomyces avermectilies* (b) *Saccharopolyspora spinosa*
   (c) Both (a) and (b) (d) None of these

95. Avermectins group includes________insecticides.

(a) Abamectin (b) Emamectin benzoate

(c) Spinosad (d) Both (a) and (b)

96. Spinolyns group of insecticide derive insecticide from________actinomycetes.

(a) *Streptomyces avermectilies* (b) *Saccharopolyspora spinosa*

(c) Both (d) None of these

97. Chlorfenapyr is belong to________insecticide group.

(a) Pyrrole (b) O. P.

(c) Avermectins (d) None of these

98. In which year Central Directorate of Plant Protection and Quarantine established?

(a) 1945 (b) 1946

(c) 1958 (d) 1957

99. Central Insecticide Board is located at_________.

(a) Faridabad (b) Ahmadabad

(c) Mumbai (d) Ranchi

100. Bombay Agricultural Pest and Disease Act was established in________year.

(a) 1947 (b) 1958

(c) 1928 (d) 1961

101. Insecticide Act was established in________year.

(a) 1945 (b) 1935

(c) 1968 (d) 1957

102. The chemical which are control to mites is called as________.

(a) Avicides (b) Rodenticides

(c) Acaricide (d) Molluscicides

103. The chemical which are control against snail is called as________.

(a) Acaricide (b) Molluscicide

(c) Nematicide (d) Avicide

104. Name the Book written by Miss. Rachel Carson.

(a) Silent spring (b) Applied Entomology

(c) Economic Entomology (d) Spring

105. ________is the stomach poison type of insecticide.

(a) Cypermethrin (b) Endosulfan

(c) Led arsenate (d) Rotenon

106. Which one of the insecticide is contact insecticide?

(a) Endosulfan (b) Malathion

(c) Monocrotophos (d) Dimethoate

107. Which one of the insecticide having fumigating action?

(a) Monocrotophos (b) Malathion

(c) Dimethoate (d) DDVP

108. Dimethoate is which type of insecticide?

(a) Respiratory (b) Physical

(c) Translaminar (d) Systemic

109. Phorate is available in which formulation?

(a) Granules (b) Dust

(c) Fumigant (d) Emulsifiable concentrate

110. Zinc phosphide is which type of poison?

(a) Rodenticide (b) Molluscicide

(c) Acaricide (d) Nematacide

111. Pesticide remain on the plant in toxic form is called as________.

(a) Poison stage (b) Residue

(c) Without toxic (d) Harmful stage

112. According to which act manufacturing of pesticide, export, or formulation making before that registration is necessary?

(a) 1968 (b) 1994

(c) 1949 (d) 1914

113. ________ is animal origin insecticide.

(a) Cartap (b) Rotenon

(c) Achook (d) Pyrathrum

114. Silent spring book was written by________.

(a) Muller (b) Kinpling

(c) R. Carson (d) Fenbrink F.

115. ________ value of insecticide is expressed in terms of mg/kg body weight of test animal.

(a) LD_{50} (b) LC_{50}

(c) LT_{50} (d) RL_{50}

116. The pesticides used against birds is called as ________.

(a) Avicides (b) Bactericides

(c) Molluscicide (d) Neemicide

117. ________ μ is particle size of dust.

(a) 110 (b) 1000

(c) 10 (d) 100

118. ________is source of Nereistoxin.

(a) *C.carnea* (b) *D. elliptica*

(c) *L. heteropoda* (d) *T. chilonis*

119. ________ in 1939 discovered the insecticidal property of DDT in 1939.

(a) Bassi (b) P. Muller

(c) E. Kinpling (d) E. Stainhus

120. ________organochlorine insecticides possess acaricidal property.

(a) BHC (b) DDT

(c) Dicofol (d) Dimthoate

121. Pest control act was passed in year________.

(a) 1689 (b) 1968

(c) 1947 (d) 1974

122. ________ is cronic rodenticide.

(a) Warfarin (b) Aluminium phosphide

(c) Zinc phosphide (d) Bromadioline

123. The amount of pesticide remaining on plant at harvest is called as________.

(a) Residue (b) MRL

(c) Eminent (d) Half life

124. ________ is organophosphates insecticide.

(a) DDT (b) BHC

(c) Malathion (d) Carbaryl

125. ________ is example of synthetic pyrethroid.

(a) DDT (b) Deltamethrine

(c) Malathion (d) Carbaryl

126. ________ is example of chitin synthesis inhibitor.

(a) DDT (b) Deltamethrine

(c) Diflubenzuron (d) All of these

Answer Key

1–c	22–c	43–b	64–a	85–a	106–b
2–c	23–a	44–a	65–d	86–b	107–d
3–a	24–a	45–d	66–c	87-a	108–d
4–a	25–a	46–a	67–b	88–c	109–a
5–a	26–b	47–c	68–a	89–c	110–a
6–c	27–d	48–c	69–a	90–d	111–b
7–b	28–b	49–b	70–a	91–d	112–a
8–b	29–a	50–d	71–d	92–a	113–a
9–d	30–d	51–a	72–a	93–a	114–c
10–d	31–a	52–c	73–d	94–a	115–a
11–b	32–d	53–d	74–a	95–d	116–a
12–a	33–d	54–a	75–d	96–c	117–d
13–c	34–d	55–d	76–d	97–a	118–c
14–a	35–a	56–a	77–c	98–b	119–b
15–a	36–a	57–d	78–a	99–a	120–c
16–b	37–b	58–a	79–a	100–a	121–c
17–d	38–d	59–c	80–a	101–c	122–a
18–c	39–c	60–b	81–a	102–c	123–a
19–a	40–c	61–a	82–c	103–b	124–c
20–a	41–a	62–a	83–b	104–a	125–b
21–c	42–a	63–a	84–b	105–c	126–c

Chapter 11
Host Plant Resistance

1. Father of HPR (host plant resistant): R. H. Painter (ICAR-NET 2013)
2. 1st ancient book containing methods of plant protection: Vrikshayurveda, author- Surapala.
3. Howlett H.B. appointed as first pathological Entomologist in 1907 at IARI.
4. Leucerne resistance to spotted aphid is associated with: High saponin in leaves.
5. Phenological asynchrony: Lack of synchrony between the life cycle of a pest and the appropriate stage of its host plant.
6. Attractant: a chemical that causes an insect to make oriented movements towards the source of the stimulus.
7. Repellent: a chemical that causes an insect to make oriented movements away from the source.
8. Feeding or oviposition stimulant: a chemical that elicits feeding or oviposition. "Feeding stimulant" is synonymous with "phagostimulant".
9. Deterrent: a chemical that inhibits feeding or oviposition.
10. Abaxial (lower surface): directed away from axis; *e.g.*, lower surface of normal leaves.
11. Adaxial (upper surface): directed towards axis; *e.g.* the upper surface of normal leaves.
12. Airspeed: speed of a moving object, *e.g.*, a flying insect, through the air.
13. 'Under hill' wheat reported some resistance to Hessian fly (1792).

14. 'Winter Majestin' apples reported to be resistant to wooly aphid.
15. C. V. Riely grafted European grapes on American rootstocks resistant to grape phylloxera (introduced from N. America).
16. At Kansas state university R. H. Painter is widely recognised as the "father of Host plant resistance".
17. Release of 'Atlas Sorgo' Sorghum resistant chinch bugs (1928).
18. Release of 'Kawvale' wheat resistant to Hessian fly (1931).
19. First course in plant resistance to insect taught by painter (1937).
20. Snelling defined plant resistance as those characters that enable a plant to avoid, tolerate or recover from attacks of insects under conditions that would cause greater injury to other plants of the same species (1941).
21. Painter described plant resistance as the relative amount of heritable qualities that influence the ultimate degree of damage done by the insect (1951).
22. Painters's book, "Insect Resistance to Crop Plants" published (1951).
23. Beck defined resistance as collective heritable characteristics by which a plant species, race, clone or individual may reduce the probability of successful utilization of that plant as a host by an insect species, race, biotype or individual (1965).
24. Nonpreference (antixenosis): host plant mechanism which includes morphological, physical or structural qualities that interfere with insect behaviour such as mating, oviposition feeding and ingestion. Deters pest before colonization.
25. Antibiosis: Interferes with pest attack due to chemical constituent of plant -Pest has reduced survival, fecundity, reproduction, etc.
26. Tolerance: Tolerance refers to the ability of the host plant to with stand an insect population sufficient to damage severely the susceptible plants.
27. Pyramiding: combine multiple resistance genes (against one pest) in the hybrid or cultivar.
28. Cowpea trypsin inhibitor gene (CpTi) was introduced into tobacco, potato, and oilseed rape for developing transgenic plants.
29. Secondary Metabolites: They do not play a role in photosynthesis, growth, or respiration.
30. Examples of commercially useful plant secondary metabolites as pesticides: nicotine and the related compound nornicotine, the pyrethrins, and rotenone.

31. Signal transduction: The ability of a cell to change behaviour in response to a receptor-ligand interaction.
32. Stress hormones: Jasmonic acid, Salicylic acid, Oligosaccharines.
33. Methyl jasmonate (MeJA) is a substance used in plant defense and many diverse developmental pathways such as seed germination, root growth, flowering, fruit ripening and senescense.
34. Methyl jasmonate is derived from jasmonic acid and the reaction is catalyzed by S-adenosyl-L-methionine: jasmonic acid carboxyl methyltransferase.
35. DIMBOA: maize (resistance factor) First generation.
36. Theory of chemosensory bases of host selection was proposed by Dethier.
37. Host plant selection process: Host habitat finding – Host finding – Host recognition – Host acceptance – Host suitability.
38. The number of biotype in rice brown plant hopper: 5.
39. The number of biotypes in rice gall midge: 13.
40. The occurrence of benzyl alcohols in minute quantities in barley was considered to be primary cause of resistance to green bug.
41. The term antixenosis proposed by Korgan and Ortman.
42. Token stimuli theory: Frankel.
43. Duel discrimination theory: Kenedy and Booth.
44. Nutrient imbalance theory: House.
45. Arka Kusumkar and Arka Shirish varieties of brinjal are resistant to Hadda beetle.
46. Phule Yashoda, M-35-1, Swati (sorghum varieties) are resistant to shoot fly (*Antherigona soccata*).
47. Golden delicious, Northern spy and Norton stocks 778, 779 are resistant to apple wooly aphid.
48. A gene transfer method involves the bombardment of intact plant with very small DNA coated sphere of tungsten or gold: biolistic (particle) Bombardment or gene gun method.
49. Insect resistant GM plants are under field testing in India: Okra, Cotton, Brinjal, Rice, Tomato, Pigeon pea, Maize, Cabbage, Cauliflower.
50. Hot spot: Hot spot locations where the pest populations are known to occur naturally and regularly at levels that often results in severe damage, are idea to test large numbers of germplasm accessions.

51. Hot spot location for *Chilo partellus* are Hisar in Haryana and Warangal in Andhra Pradesh.

52. Hot spot locations for sorghum midge are Dharwad, Bhavanisagar and Pantanagar in India.

53. Green and Ryan reported that insect herbivory induce the accumulation of proteinase inhibitors (PIs) in tomato leaves.

54. Bowman- Brik Type PI-small polypeptidase, found in legume seed.

55. Phytoncide: antimicrobial allelochemic volatile organic compounds derived from plants.

56. Phytoncide word means ' exterminated by the plant'.

57. Phytoncide word coined by Dr. Boris P. Tokin a Russian biochemist.

58. DIMBOA ia an aglucone [2,4-dihydroxy-7-methoxy-2-H benzoxazin-3 (4H)-one].

59. When normal healthy foliage of maize is mechanically damaged, the glucoside 2-O-glucosyl-4-hydroxy-1, 4-benxoxazin-3-one is enzymatically converted to DIMBOA works as deterrence for feeding by European corn borer.

60. Cloning of the Mi-gene from wild tomato (*Lycopersicum peruvianum*) has given an opportunity to control root-knot nematode and potato peach aphids.

61. Lectins are proteins having affinity for specific carbohydrate moieties.

62. Lectins bind to glycoproteins in the peritrophic matrix lining the insect midgut to disrupt digestive process and nutrient assimilation.

63. Wheat germ agglutinin, pea lectin, jacalin and rice lectin have been expressed in plant like tobacco, maize and potato mainly against aphids.

64. Antibiosis will put high pressure on biotype development.

65. Antixenosis will put little pressure on biotype development.

66. Tolerance will put no pressure on biotype development.

67. The rice variety mudgo resistance to BPH due to nutritional deficiency.

68. *Bt* resistance occurs through loss of cadherin, which has important roles in cell adhesion, ensuring that cells within tissues are bound together. This mechanism is found, for instance, in *Bt*-resistant diamondback moth (*Plutella xylostella*).

69. GMO: Genetically Modified Organism.

70. DIMBOA (Dihydroxy methyl benzoxazin) imparts resistance against: European corn borer, *Ostrinia nubilalis.*

71. Gossypol (Polyphenol) imparts resistance against: *Helicoverpa armigera* (American bollworm).

72. Sinigrin imparts resistance against: Aphids, *Myzus persicae.*

73. Cucurbitacin imparts resistance against: Cucurbit fruit flies.

74. Salicylic acid imparts resistance against: Rice stem borer.

75. The 1st transformation of delta-endotoxin gene from Bt was done in tobacco (Belgium, 1985).

76. Cumbu (HB1) and Hybrid sorghum (CSH 1) favour shoot flies and gall midges.

77. Non preference mechanisum by host plant is also known as: Antixenosis (ICAR-NET 2013).

78. Abadita (cotton) resistant to American bollworm.

79. M 35-1, Swati varieties of sorghum resistant to shoot fly.

Examples of Secondary Metabolites Associated with Specific Tissues or Structures

Tissue	*Chemical*	*Plants*
Trichomes	monoterpenes	Lamiaceae
Trichomes	sesquiterpenes	Solanaceae
Trichomes	Flavonoid glycosides	Solanaceae
Wax	triterpenes	Asclepiadaceae
Wax	phenolics	Rosaceae
Epidermis	quinolizidine alkaloids	*Genista*
Cell walls	tannins	many trees
Dead cells	tannins	woody plants
Vacuoles	cyanogenic glycosides	many
Vacuoles	alkaloids	many
Oil glands	furanocoumarins	*Citrus*
Oil glands	sesquiterpenes	*Gossypium*
Latex	di- and triterpenes	Euphorbiaceae
Latex	lactones sesquiterpene	Asteraceae

Tissue	*Chemical*	*Plants*
Latex	alkaloids	Euphorbiaceae
Resin canals	diterpenes	Gymnosperms
Seeds	non-protein amino acids	Legumes
Seed coat	furanocoumarins	*Pastinaca*
Bark	quinine	*Cinchona*

Some Examples of Transgenic Plants Expressing Gene Encoding Proteinase Inhibitors, α-amylase Inhibitors and Lectins

Crop	*Gene*	*Target Crop*
Tobacco	Cow pea serine PI	Tobacco budworm
Tobacco	Potato serine PI	Tobacco horn worm
Tobacco	Horn worm PI	White fly
Rice	Cow pea serine PI	Stem borer
Rice	Snowdrop lectin	Brown plant hopper
Potato	Oryzacyatatin	Potato beetle
Potato	Cow pea serine PI	Lacanobia
Potato	Snowdrop lectin	Potato aphids
Pea	Bean α-AI	Bruchids

Chapter 12
Economic Entomology

1. Floral fidelity: A bee visits same species of plant for pollen/nectar collection until exhausted.
2. Honey is hygroscopic. If exposed to air it absorbs moisture.
3. Specific gravity of pure honey is 1.35 - 1.44 g/cc.
4. Oil palm pollinating weevil: *Elacidobins kamerunicus* (Curculionidae : Coleoptera).
5. Entomophobia: an unnatural and unreasonable fear of arthropods.
6. Oligogyny: The presence of two or more functional queens in a nest.
7. Bivoltine silkworm: completes two life cycles in a year (alternating cycles of hibernating and non hibernating eggs).
8. Biovoltine race of silkworm (in India): Kalimpong, KA, NB7, NBD2, NB-18.
9. Bivoltine breeds of silkworm (Tamil Nadu): SK3, SK4, SK6, SK7 and YB.
10. One worker honey bee only makes 1/12th a teaspoon of honey in their life time.
11. Worker bees secrete wax when they are 14-18 days old.
12. Bave is the silk thread spun by a silkworm.
13. Chandrika is a bamboo montage used for making silkworm for spinning cocoon.
14. Croissure as a technique for twisting raw silk over itself during a process of reeling.
15. Deflossing: method of removing the floss from the cocoons.

16. Poisonous pollen for honey bees: *Aesculus hippocastanum*, *Rhododendron* sp., *Tillia* sp., *Cyrilla* sp.
17. Parasite of a silkworm: *Strumia caricaric.*
18. *Apis florea* species of honey bee produces the sweetest honey.
19. Artificial honey bee hive was prepared first time by Langstroth.
20. Honey robber is also known as *Acherontia styx*.
21. Moulting occurs in a silkworm: 4 times.
22. Central silk Research Board is situated at: Baharampur (WB).
23. Silk research centre in Uttaranchal is situated at: Dehradun.
24. The female lac insect is three times larger than male.
25. The organs absent in female lac insect: eyes, legs, wings.
26. Adult lac male insect survives only for 3-4 days.
27. Colour of honey is measured using a "Pfund grader" (named after the inventor Dr. Pfund).
28. Honey bee venom contains: Melitinin (IARI Ph. D. 2011).
29. Honey bee species whose nests consist of multiple combs: *Apis cerana, Apis mellifera, Apis nuluensis, Apis koschevnikovi* and *Apis nigrocincta.*
30. Honey bee species whose nests are single combs: *Apis florea, Apis dorsata, Apis andreniformis, Apis binghami, Apis laboriosa* and *Apis breviligula* (IARI Ph. D. 2011).
31. The immature lac is known as Ari lac.
32. Lac insect reproduction: Ovoviviparous.
33. One antioxidant called 'pinocembrin' is only found in honey.
34. Bee venom (chemical composition): melittin, phospholipose A, hyaluronidase, histamine and Acid phosphatise.
35. Propolis is known as 'bee glue'.
36. Floss is low grade silk in the outer part of cocoon.
37. Croissure is a technique for twisting row silk over itself during the process of realing.
38. "Pigmentation stage" of silk worm: on 8th day colour of egg become light with distinct black spot.
39. Pebrin disease of silkworm: pepper like black spot (ICAR-SRF, 2011).

40. Sericulture is derived from French word (seris- silk).
41. Clitellum: thich band like stracture of earthworm (segment 14 to 16).
42. Clitellum function: secretion of mucus, egg case and cacoon.
43. Earthworm lacks distinct head and scene organ (eyes).
44. Earthworm gizzard lise in 8th segment of body.
45. South Indian earthworm: *Megascolex maurtii.*
46. *Nereis* belongs to class polychaeta.
47. Fig wasp (*Blastophaga spp.*) belongs to family Agaonidae.
48. Real product of honey bees: bee wax.
49. Seribiotech Research Laboratory located at Bangalore.
50. Central Sericulture Germplasm Research Centre: Hosur (Tamil Nadu).
51. Honey bee worker: sterile females.
52. Semi domesticated silkworm: Muga (ICAR-SRF, 2013).
53. Lac is mainly secreted by : females.
54. Silk gland : ectodermal (ICAR-NET 2013).
55. Muscardine disease of silkworm: fungal disease (ICAR-NET 2013).
56. Resin is secreted by glands of lac insect. (ICAR-NET 2013).
57. Blister beetles (family Meloidae) contain relatively high concentrations of cantharadin, a compound with notable effects on the vertebrate urogenital system (aphrodisiac).
58. Maggot therapy: sterile-reared fly larvae for cleaning necrotic tissue from deep wounds.
59. Muscardine disease of silkworm is caused by *Beauveria bassiana.*
60. Open pan cooking method for stifling of silkworm has been developed by CSR and TI.
61. A single silkworm caterpillar is capable of producing 650 to 1300 meters of silk thread.
62. Haluthonde/Grasserie disease of silkworm caused by nuclear polyhedrosis virus.
63. Host of lac insect (kusum): *Schelichera* spp. (ICAR-NET 2013).
64. The insect used for crime detection: house fly (ICAR-NET 2013).
65. Insect has been used to study genetics: *Drosophila melanogaster* (ICAR-NET 2013).

Fill in the Blank Type Questions with Answers

1. **The family of honey bee __________.**
   Ans: Apidae

2. **__________ is introduced honey bee in India.**
   Ans: A. mellifera

3. **__________ is commercially important honey bee.**
   Ans: A. mellifera

4. **__________ Honey bee is irritable and ferocious nature.**
   Ans: A. dorsata

5. **__________honey bees build a many parallel combs.**
   Ans: A. cerana indica

6. **__________ is little and smallest bee found only in plains.**
   Ans: A. florea

7. **Yeilding potential of Italian honey bee is__________kg/year.**
   Ans: 40-170

8. **Italian honey bee, *A. mellifera* first time introduced in India in the year __________.**
   Ans: 1960

9. **Dammer bee, *Trigona iridipennis* made a comb of __________which is mixture of earth and wax.**
   Ans: Cerumen

10. **In honey bee fertilized eggs developed into__________.**
    Ans: Female workers

11. **Ovipositor is modified as stinging organ in__________of honey bees.**
    Ans: Workers

12. **Crop of honey bee called__________.**
    Ans: Honey stomach

13. **Honey bee saliva have__________enzyme.**
    Ans: Invertase

14. **Royal jelly is secreted by__________gland.**
    Ans: pharengeal

15. **Fertilised eggs of queen honey bee developed in to __________.**
    Ans: Worker

16. **Dumbel shaped mandibles found in __________insect.**
Ans: Honey bee

17. **In honey bee the ventral nerve cord have__________number of ganglions.**
Ans: 7

18. **Royal jelly produced by __________cast of honey bee.**
Ans: Worker

19. **Wax gland present on __________abdominal segments in honey bee.**
Ans: 4-7

20. **__________is venom of bee sting.**
Ans: Formic acid

21. **Wax plates, pollen basket are absent in __________ bee.**
Ans: Queen

22. **__________ inhibit the production of new queen, which is secreted by queen.**
Ans: Queen substance

23. **Queen honey bee lays__________ egg per day.**
Ans: 500

24. **The life cycle of worker bee is__________ weeks.**
Ans: 6

25. **Scent glands are present on __________ abdominal segments of worker bees.**
Ans: 5-6

26. **__________dance indicate the direction of food source in honey bee.**
Ans: Wag tail

27. **Trehalose is __________ sugar.**
Ans: Disaccharide

28. **Language of communication in honey bee through wag tail dance was first searched by __________.**
Ans: Karl Frisch

29. **Swarming in honey bee done in __________ month.**
Ans: Nov.–March

30. **In __________ year, India started the work on scientific beekeeping.**
Ans: 1907

31. **In 1851 movable frame hive was developed by__________.**
Ans: Langstroth

32. __________ hive was popular and largely used in south India.

Ans: Newton

33. In artificial hive, honey is stored at __________ chamber.

Ans: Super

34. __________ plant provide large quantity of nectar to bees.

Ans: Tamrind

35. __________ per cent moisture present in honey.

Ans: 20

36. __________ is a sex pheromone present in honey bee.

Ans: Queen substance

37. Unripe honey or honey in unsealed cells known as __________.

Ans: Green honey

38. Specific gravity of wax is __________.

Ans: 0.95 g/cc

39. Which of the following are the enemies of bees.

Ans: Wax moth, Yellow jucket wasp andWax beetle

40. *Galleria mellonella* is a pest of __________.

Ans: Honey bee

41. Bald brood disease caused by __________ to honey bee.

Ans: Achroia grisella

42. Thai sac brood disease introduced in India through __________ bees.

Ans: Apis mellifera

43. Brown and putrefied pupae giving fishy odour in bees are the symptoms of __________ disease.

Ans: American foul brood

44. __________: ectoprasitic mites infects the honey bee

Ans: Varroa jacobsoni

45. __________ is method of reproduction and perpetuation of species in honey bee.

Ans: Swarming

46. Isle of Wight or Acarine disease is caused by __________.

Ans: Acarapis woodi

47. Silk is produced in silkworm by__________gland.

Ans: Labial

48. Silk have__________ and __________protein found about 75 per cent and 25 per cent respectively.

Ans: Fibroin and sericine

49. Hibernation occurs in which stage of *bombyx mori*.

Ans: Egg

50. __________race of *Bombyx mori* produces superior quality silk.

Ans: Univoltine

51. Circular basket required to attach cocoon in sericulture is known as.

Ans: Chandrika

52. Rearing of 1st and 2nd instar of silkworm upto 2nd moult is known a__________rearing.

Ans: Chawki

53. Silkworm feeds__________gm of leaves during its life cycle.

Ans: 90-100

54. Killing of pupa inside the cocoon is known as__________.

Ans: Stifling

55. Muscaridine in silkworm is__________disease.

Ans: Fungal

56. __________is endoparasitie of mulberry silkworm.

Ans: Exorista sorbillans

57. Host plant of Eri silkworm is__________.

Ans: Caster

58. Eri silkworm commercially reared in__________.

Ans: Assam

59. Colour of Eri silkworm silk is__________.

Ans: Brick-red or white

60. Muga silk worm commonly reared in__________state.

Ans: Assam

61. __________silkworm is not domesticated.

Ans: Antheraea mylitta

62. Shorea and Ber are important host plant of__________.

Ans: Tasar silkworm

63. ________silkworm is semi domesticated.

Ans: Muga

64. The host plants for *Antheraea assamens* is________.

Ans: Machilus bombycina and *Litsaea polyantha*

65. Fibroin is elastic and tough protein present________ per cent of silk threat on inner side.

Ans: 75

66. Weight of 900gm long silk filament expressed in gram is________.

Ans: Denier

67. Family of *Morus india* (Mulberry) is________.

Ans: Moracea

68. Silk of________silkworm is unreelable.

Ans: Eri

69. ________is protozoal disease of silkworm.

Ans: Pebrine

70. *Beauveria bassiana* produces________muscaredine in silkworm.

Ans: White

71. Muga silkworm lays eggs on dry twigs known as________.

Ans: Kharika

72. Transfer of hatched larvae to host leaves is known as________.

Ans: Brushing

73. ________is known as poor man's silk.

Ans: Erisilk

74. Central Tasar Research Station located at________.

Ans: Ranchi

75. Tassar, Eri and Muga silkworms belong to family________.

Ans: Saturniidae

76. ________is a major producer of lac in lac insect.

Ans: Female

77. Granular lac is known as________.

Ans: Seed lac

78. Lac is marketed as thin sheet is known as ________.

Ans: Shellac

79. __________ is enemy of lac insect.

Ans: Eublemma amabilis, Holocera pulverea and *Chrysop*a

80. __________country is the largest producer of lac.

Ans: India

81. __________type of reproduction observed in lac insect.

Ans: Oviparous

82. __________not host plant of lac insect.

Ans: Morus indica

83. *Shorea oleosa* is host plant for__________insect.

Ans: Lac insect

84. Baisakhi crop harvested after __________months.

Ans: 8 months

85. Scraped raw lac is also known as __________.

Ans: Stick lac

86. Molemma have __________ per cent shellac.

Ans: 70 per cent

87. __________is used for inoculation of lac insect.

Ans: Seed lac

88. __________ is species of Indian Honey bee.

Ans: Apis cerena

89. Indian Lac Research Institute located at___________.

Ans: Ranchi

90. National Center for Agricultural Important Insect (NCAII) located at___________.

Ans: Bangalore

91. ___________country is larger producer of lac which meet 60 per cent of world's requirement.

Ans: India

92. ___________ plants are used for rearing Tassar silk worm.

Ans: Mulberry

93. ___________ Varity of Mulberry used in large scale in Maharashtra.

Ans: V-1

94. __________ host plants used for lac production.

Ans: Ber

95. __________ (country) produces all four types of silk.

Ans: India

96. National Center for Agricultural Important microorganisms (NCAIM) located at__________.

Ans: Mau Bhajan (U.P.)

Chapter 13
Pheromone

1. Pheromone (beneficial to receiver): kairomone (ICAR-NET 2013).
2. The most potential traps are pheromone trap.
3. COCO-Trapp™: trap patented and introduced by PCI (Pest control of India).
4. COCO-Trapp™-For specially: tapping Red palm weevil and Rhinocerous beetle.
5. Wota-T ™, Del-Ta™: traps by PCI.
6. Fight-T™: yellow bowl with entry hole to dome.
7. Fight-T™: For trapping melon fly and oriental fruit fly.
8. Roban ready to use wax block bait for rodents.
9. Pest-O-Flash®: (Electrical flying insect killer)- India's 1st electrical flying insect control system.
10. Trubble gum®: non toxic Glue traps for rat and mice.
11. Pheromones used for monitoring purpose- dose not requires registration.
12. Male produces pheromone: Red palm weevil, Cotton boll worm, Banana corm weevil.
13. P B Rope-L used for mating destruption of pink boll worm.
14. Alarm pheromone in case of ants: formic acid.
15. Alarm pheromone in case of Termite soldiers: Aldehydes.
16. Alarm pheromone in case of Aphids: Terpenes.

17. L- Lactic acid and CO_2 are found lures for *Aedes* sp.
18. Mosquito repellent: DEET [N, N, Diethyl m-tolumide].
19. Chemical name of Bombykol: 10,12 – hexadeca dienol.
20. Bombykol is the natural ligand for a pheromone binding protein, BmorPBP, which escorts the pheromone to the pheromone receptor.
21. Epidietic pheromone: the compounds functional in the regulation of population density (by dispersion of organisms).
22. Pheromone lure (lucinlure): *Leucinodes orbanalis.*
23. Oryctalure: Rhinocerous beetle.
24. Dimethyl decaneyl is aggregation pheromone for *Tribolium confusum.*
25. Wota-T ™, Del-Ta™ traps by PCI.

Multiple Choice Questions

1. **Pheromones are __________ in nature.**
   (a) Repellent (b) Attractant
   (c) Deterrent (d) Antijuvenil
2. **The term pheromone was coined by__________in space 1959.**
   (a) Karlson and Butenandt (b) Helmate
   (c) David (d) Pradhan
3. **Sex pheromones are mostly released by_________insect of particular species.**
   (a) Female (b) Male
   (c) Both male and female (d) None of these
4. **_________is best chitin synthesis inhibitor.**
   (a) Urea (b) Thiourea
   (c) Diflubenzuron (d) Lufenuron
5. **CSI (chitin synthesis inhibitors) are mostly_________ poison.**
   (a) Stomach (b) Respiratory
   (c) Nerve (d) None of these
6. **_________are examples of chitin synthesis inhibitor.**
   (a) Buprofezin (b) Novaluron
   (c) Lufenuron (d) All of these

7. **_________is naturally occurring CSI.**
   (a) Plumbagin (b) Buprofenzin
   (c) Both (d) None off these

8. **_________is an sclerotization inhibitor developed by Monsanto.**
   (a) MON 0585 (b) MON 112
   (c) MON 0240 (d) MON 1818

9. **_________are JH analogues used in pest control.**
   (a) Methoprene (b) Hydroprene
   (c) Fenoxycarb (d) All of these

10. **_________is the moulting hormone in insect.**
    (a) Methoprene (b) Hydroprene
    (c) Ecdysone (d) Fenoxycarb

11. **Pyriproxyfen belongs to_________anologues.**
    (a) MH (b) PTTH
    (c) JH (d) None of these

12. **Which of the following lure used to attractant *Bactrocera dorsalis*.**
    (a) Helilure (b) Cue lure
    (c) Hexalure (d) Trimedlure

13. **_________is sex pheromone of *Bombyx mori*.**
    (a) Gossyplure (b) Bombykol
    (c) Gyplure (d) Hexalure

14. **_________is sex pheromone for *Lymantria dispar*.**
    (a) Gossyplure (b) Gyplure
    (c) Hexalure (d) Vitlure

15. **_________lure is used to control cabbage looper.**
    (a) Vitlure (b) Gossyplure
    (c) Hexalure (d) Looplure

16. **_________lure is used to attract spotted bollworm.**
    (a) Ervitlure (b) Vitlure
    (c) Spodolure (d) Both (a) and (b)

17. **The β-farnesene act as_________pheromone in aphids.**
    (a) Alarm (b) Aggrigation
    (c) Trail (d) Sex

18. Kairomone are benefited to________Insect.

(a) Sender (b) Receiver

(c) Both Receiver and Sender (d) None of these

19. Which of the following semiochemicals acting in interspecific.

(a) Kairomone (b) Synomone

(c) Allomone (d) All of these

Answer Key

1–b	**5–a**	**8–a**	**11–c**	**14–b**	**17–a**
2–a	**6–d**	**9–d**	**12–b**	**15–d**	**18–b**
3–a	**7–a**	**10–c**	**13–b**	**16–c**	**19–d**
4–c					

Chapter 14
Acarology

1. Adult mite possesses four pairs of legs.
2. The chelicherae are found in Arachnida.
3. The body is divided into cephalothorax and abdomen in Arachnida.
4. In mites respiration takes place in book lungs.
5. The mite *Acarus siro* damages store grains.
6. Spider mites belong to family Tetranychidae.
7. Eriophyid mites possess two pairs of legs.
8. Mite, *Polyphagotarsonemus latus* causes 'tambera disease' in potato.
9. Mite, *Polyphagotarsonemus latus* causes 'murda disease' in chilly.
10. Eriophyidae was first identified in Guerrero states of Mexico during 1960. It was first described by (Keifer, 1965).
11. Coconut mite: *Aceria guerreronis* Keifer (ICAR-NET, 2012).
12. Hard ticks ' questing position': when seeking a host, hard ticks usually climb on vegetation and adopt a posture known as questing.
13. Isle of Wight disease of honey bee due to mite (*Acarapis woodi*). (ICAR-NET 2013).

Important Species of Mites

Common name		Scientific name
Brown wheat mite		*Petrobia latens*
Cheese mite		*Tyrolichus casei*
Chicken mite		*Dermanyssus gallinae*
Citrus leaf mite		*Eutetranychus banksi*
Coconut mite		*Aceria guerreonis*
Cotton mite		*Teranychus macfarlanei*
Cucurbitaceous mite		*Teranychus cucurbitae*
Grain mite		*Acarus siro Linnaeus*
Honey bee mite		*Varroa* sp.
Honey bee tracheal mite		*Acarapis woodi*
House Dust mite		*Dermatophagoides* sp.
Itch mite/Scabies mite		*Sarcoptes scabiei*
Jasmine eriophyid mite		*Aceria jasmini*
Mange mite (dog)		*Demodex canis*
Mango mite		*Aceria mangiferae*
Mold mite		*Tyrophagus putrescentiae*
Paddy mite		*Oligonychus oryzae*
Sorghum mite		*Oligonychus indicus*
Straw Itch mite		*Pyemotes tritici*
Sugarcane mite		*Schizotetranychus andropogoni*
Two spotted spider mite		*Teranychus urticae*
Two- spotted spidermite		*Tetranychus urticae*
Yellow mite/broad mite		*Polyphagotarsonemus latus*
Tea mite	Red spider mite	*Oligonychus coffeae*
	Scarlet mite	*Brevipalpus californicus*
	Purple mite	*Calacarus carinatus*
	Pink mite	*Acaphylla theae*
	Yellow mite	*Phytotarsonemus latus*

Chapter 15

Veterinary and Medical Entomology

1. Mosquitoes belong to family (Culicidae: Diptera).
2. Mosquitoes belong to sub order Nematocera.
3. Human disease transmitted by mosquitoes: malaria, yellow fever, dengue fever, filariasis and encephalitis.
4. Mosquitoes belong to family (Culicidae: Diptera).
5. Mosquitoes belong to sub order: Nematocera.
6. Human disease transmitted by mosquitoes: malaria, yellow fever, dengue fever, filariasis and encephalitis.
7. Animal diseases transmitted by mosquitoes: Fowl pox of poultry, myxomatosis of rabbit, Rift valley fever of sheep, encephalitis of horse and birds, heartworm of dog.
8. Mosquitos' only females are blood suckers; males are vegetarians (nectar and juice of fruit).
9. *Culex pipens* and *Culex molestus*: autogenous (*i.e.* oviposits without the need of blood meal).
10. Anopheline mosquitoes prefer to feed at twilight and are inactive in bright day light and night.
11. Black, blue and red are most attractive colour for Aedes mosquitoes, white and yellow the least.

12. Mosquito whine: The sound comes from the vibrating wings or produced by the tarsi of leg rubbed against the wings.
13. Anopheline and culicine oviposition in: water.
14. Aedes (oviposition): shaded localities on ground with intermittent flooding.
15. Non blood sucking families of mosquito: Dixinae, Chaoborinae, Toxorhynchitinae.
16. Culex mosquito -egg- Cigar shaped, laid in rafts and no airfloat.
17. Aedes mosquito -egg- Eliptical, laid singly and no airfloat.
18. Anopheles mosquito -egg- Boat shaped, laid singly and airfloat present.
19. Anopheles mosquito (larva): respiratory siphon absent.
20. Culex and Adeges mosquito pupa- colourless.
21. Anopheles mosquito pupa- green in colour.
22. The resting posture of anopheles mosquito: Inclined, head down at 45°C angle.
23. Anopheles mosquito wings: spotted.
24. The mosquito larva breaks out on the anterior end of the egg by using a sharp projection on its head: Egg breaker.
25. Fishes feed on mosquito (brackish breeding ground): kill fish (fundulus).
26. Fishes feed on mosquito (freshwater breeding ground): *Gumbusia affinis.*
27. The genus name *Gambusia* is derived from the Cuban Spanish term *gambusino*, meaning "useless".
28. Western mosquitofish: (*Gambusia affinis*); Eastern mosquitofish: (*G. holbrooki*).
29. In India, a local species *Aplochelis panchax* feeds on mosquitoes.
30. Plasmodium: a causative agent for malaria.
31. *Plasmodium vivax*: (tertian or vivax malaria): fever after 48 hrs or every 3rd day.
32. *Plasmodium falviparum*: (malignant tertian malaria): fever after 24-48 hrs.
33. *Plasmodium falviparum*: (quartan malaria).
34. *Plasmodium ovale*: (mild begin malaria / ovale malaria): fever after 3rd day.
35. Ronald Ross was awarded Nobel Prize in 1902.
36. The housefly belongs to suborder Cyclorrhapa.

37. House fly -*Musca nebulo*: tropical region species.
38. House fly -*Musca domestica*: temperate region species.
39. House fly transmits disease: typhoid, dysentery, cholera, yaws, T.B., leprosy, trachoma, anthrax, gonorrhoea.
40. Eye flies (*Siphunculina funicola*) belongs to family Chioropidae.
41. Lice: *Pediculus humanus corporis* (body lice).
42. Head lice: *Pediculus humanus capitis.*
43. Crab/pubic louse: *Pthirus pubis.*
44. Bed bug belongs to suborder: Heteroptera.
45. Bed bugs can also be detected by their characteristic smell of rotting raspberries.
46. *Cimex hemipterans* (= rotundus): tropical, also called as oriental/Indian bed bug.
47. *Cimex lectularius:* temperate region bed bug.
48. Bed bugs reproduce by *traumatic insemination*, also known as *hypodermic insemination.*
49. Traumatic insemination: Instead of using the genital tract, the male pierces the female's abdomen with his genitalia and injects his sperm through a wound in the abdominal cavity. The sperm migrates through her abdominal fluid until it reaches her ovaries.
50. The female bed bug lays approximately 5 eggs in one day and about 500 during her lifetime.
51. Carrion disease (orya fever): transmitted by sand fly (*Phlebotomus noguchi*).
52. Eye gnats/Eye flies for their habit of being attracted to eyes. They feed on lachrymal secretions and other body fluids of various animals, including humans.
53. Sand fly genera belong to Phlebotominae subfamily are the primary vectors of leishmaniasis and pappataci fever.
54. Paragonimiasis: disease caused by the trematode *Paragonimus westermani* by eating raw or insufficiently cooked crustaceans which act as intermediate host.
55. Triatominae: a subfamily of Reduviidae, members are also known as conenose bugs, kissing bugs, assassin bugs, or triatomines.
56. Disease transmitted by eyeflies: ophthalmia and conjunctivitis.

57. Vector of chagas disease: *Rhodnius prolixus.*
58. Hornflies: *Haematobia irritans* (Tahinidae:Diptera): Both sexes suck blood from neck region from cattle, goats, horses, gods and sheep.
59. Yellow fever transmitted by *Aedes aegyptii.*
60. The vector control research centre is located at Pondicherry.
61. Stegomyia: old nickname of *Aedes* spp.
62. Mallophaga : bird lice (malloss- wool).
63. Malaria and yellow fever transmitted by mosquitoes.
64. *Xenopsylla cheopis* belongs to Siphonoptera order.
65. Infestation of lice is called as pediculosis.
66. Diptera have the insect more usefull to medical and veterinary importance.
67. African sleeping sickness and nagana diseases transmitted by Tsetse fly.
68. Kala Azar disease transmitted by sand fly. (ICAR-NET 2013).
69. Filariasis disease in human is transmitted by mosquito. (ICAR-NET 2013).
70. Dengue disease transmitted by aedes mosquito. (ICAR-NET 2013)

Lists of Pests of Veterinarian Importance

I. Blood Sucking Flies

Common Name	*Scientific Name*	*Family*
Order Diptera		
Sandfly	*Phlebotomous* spp.	Psycondidae
Horse fly	Tabanids	Tabanidae
Horn fly	*Haematobia (=Siphona) irritans* (L.)	Muscidae
Tsetse fly	*Glossina palpalis*	Muscidae
Black fly	*Simulium aureohirtum*	Simuliidae
Louse fly	*Hippobasca maculata* Leach	Hippoboscidae
Buffalo fly	*Siphona exigua*	Muscidae
Stable fly	*Stomoxys calcitrans*	Muscidae

II. Myiasis Flies

Common Name		Scientific Name	Family
Botflies	Horse bot	*Gastrophilus intestinalis*	Oestridae
	Sheep bot	*Oestrus ovis*	Oestridae
	Warble fly	*Hyperderma lineatum*	Oestridae
Blow flies		*Chrysomyia bezziana*	Calliphoridae
		Cochliomyia (=Callitroga) hominivorx	Calliphoridae
		Calliphora erythrocephala	Calliphoridae
		Lucilia cuprina	Calliphoridae
		Phormia regina	Calliphoridae
Flesh flies	Flesh fly	*Sarcophaga* spp.	Sarcophagidae

III. Lice

Common Name		Scientific Name	Family
Order Anoplura			
Sucking lice	Hog louse	*Haematopinus* spp.	Haematopinidae
	Blue lice of cattle	*Linognathus vituli*	Linognathidae
Order Mallophaga			
Biting lice	Shaft louse	*Menopon gallinae* (L.)	Menoponidae
	Pigeon louse	*Columbicola columbae* (L.)	Philopteridae
	Chicken body louse	*Eomenocanthus (=Menacanthus) stramieus*	Monoponidae
	Animal lice	*Bevicola* spp.	Trichodectidae
	Dog louse	*Trichodectis canis*	Trichodectidae

IV. Flea

Common Name	Scientific Name	Family
Order Siphonoptera		
	Echidnophaga gallinacea	Hectosylidae
	Ctenocephalidaes canis	Pulicidae

Disease and Vectors

Common Name of Insect Vector	*Scientific Name*	*Name of Disease*	*Causal Organism*
Sandfly	*Phlebotomus intermedius*	American leishmaniasis	*Leishmania brasiliensis*
	Phlebotomus noguchi	Carrion disease	*Bacillus anthracis*
	Phlebotomus serganti	Oriental sore	*Leishmania tropica*
	Phlebotomus argentipes	Kala azar/Dumdum fever/Black fever/Leishmaniasis	*Leishmania donovani*
House fly	*Musca* sp.	Cholera	*Vibrio cholera*
		Amoebiasis	*Entamoeba histolyca*
		Dysentery	*Shigella sonnei*
		Typhoid fever	*Salmonella typhi*
		Polimyelitis	Virus
		Tuberculosis	*Mycobacterium tuberculosis*
Black fly	*Simulium* sp.	Onchocerciasis/River blindness	*Onchocera volvulus* (nematode)
Horse fly		Anthrax	*Bacillus anthracis*
Deer flies	*Chrysops* spp.	Loasis	*Loa loa*
Mosquitoes	*Aedes aegypti*	Dengue	Virus
	Aedes aegypti	Yellow fever	Virus
	Culex sp.	Encephalitis	Virus
	Culex quinquefasciatus	Filariasis/Elephantiasis	*Wuchereria bancrofti*

Common Name of Insect Vector	Scientific Name	Name of Disease	Causal Organism
	Anopheles sp.	Malaria	*Plasmodium* sp.
Tsetsefly	*Trypanosoma gambiens*	Gambian sleeping sickness/ African trypanosomiasis	*Glossina palpalis*
	Trypanosoma rhoesiense	Rhodesian sleeping sickness/ Rhodesian trypanosomiasis	*Glossina morsitans*
Assain bug/ reduvid bug	*Rhodnius prolixus* *Triatoma infestans*	Chagas disease or American trypanosomiasis	*Trypanosoma cruzi*
Lice	*Pediculus humanus* (Human lice)	Relapsing fever	*Rickettsia typhi*
	Pediculus humanus (Human lice)	Epidemic typhus	*Borellia recurrentis*
Rat flea	*Xenopsylla*	Endemic typhus	*Rickettsia prowazeki*
	Xenopsylla cheopsis	Bubonic plague	*Yersinia pestis*

V. Arachnids

Common Name		*Scientific Name*	*Family*
Order Acarina			
Ticks	Cattle ticks	*Boophilus microplus*	Ixodidae
	Fowl ticks	*Argus persicus*	Argasidae
	Dog tick	*Rhipicephalus sanguinesus*	Ixodidae
Mites	Manage mite	*Sarcoptess scabiei*	Sarcoptidae
	Dog mite	*Demodex canis*	Demodicidae

Chapter 16
Match the Pairs

1)

1.	*Chrysoperla carnea*	a.	Dictyoptera
2.	*Coccinella septempunctata*	b.	Coleoptera
3.	Preying mantids	c.	Neuroptera
4.	*Trichogramma chilonis*	d.	Hemiptera
5.	*Cyrtorhinus lividipennis*	e.	Hymenoptera

2)

1.	Vapum	a.	Organophosphate
2.	Fenvalerate	b.	Neonicotinoid
3.	Temik	c.	Aldicarb
4.	Imidacloprid	d.	Synthetic pyrethroid
5.	Triazophos	e.	Sodium methyl dithio carbamate

3)

1.	Stem or bulb nematode	a.	Longidorus
2.	Needle nematode	b.	Aphelenchoides
3.	Foliar nematode	c.	Ditylenchus
4.	Reniform nematode	d.	Trichodorus
5.	Stubby root nematode	e.	Rotylenchus

4)

1.	*Leptocorisa acuta*	a.	Rice leaf folder
2.	*Nilaparvata lugens*	b.	Potato tuber moth
3.	*Gnorimoschema operculella*	c.	Green leaf hopper
4.	*Cnaphalocrocis medinalis*	d.	Gundhi bug
5.	*Nephotettix virescens*	e.	Brown plant hopper

5)

1.	Nosema	a.	Arhar
2.	Wax moth	b.	Protozoa
3.	Baisakhi crop	c.	*Galleria mellonella*
4.	Sac brood	d.	Lac
5.	*Heterodera cajani*	e.	Virus

6)

1.	Spring tail	a.	Odanates
2.	Air sac	b.	Honey bee
3.	Plastron respiration	c.	Termite
4.	Fermenting chamber	d.	Coleopterans
5.	Pre-hensile organ	e.	Collembola

7)

1.	Hopper burn	a.	Cotton jassid
2.	Bunchy top	b.	Gall midge
3.	Dead heart	c.	Shoot fly
4.	White earhead	d.	Stem borer
5.	Silver shoot	e.	Sugarcane top borer

8.

1.	Proventriculus	a.	Blow
2.	Resilin	b.	Aerobic
3.	Systole	c.	Rubber like protein
4.	Maggot therapy	d.	Blood circulation

9)

1.	Cowpea trypsin inhibitor	a.	Crucifers
2.	β-solanin	b.	Glycoalkaloid
3.	Leptin	c.	Gallic acid
4.	Glucosinolates	d.	*Callosobruchus chinesis*
5.	Phenol	e.	Feeding deterrent

10)

1.	*Nephotettix nigropictus*	a.	Tungro disease
2.	*Nephotettix virescens*	b.	Rice Transitory Yellowing
3.	*Nilparvata lugens*	c.	Mottling and spike disease of s'cane
4.	*Saccharicoccus sacchari*	d.	Grassy stunt
5.	*Melanaphis sacchari*	e.	Grassy shoot of s'cane

11)

1.	*Bemisia tabaci*	a.	Citrus canker
2.	*Scirtothrips dorsalis*	b.	Leaf curl of chilli
3.	*Phyllocnistis citrella*	c.	Tomato leaf curl
4.	*Nephotettix nigropictus*	d.	Groundnut rosette
5.	*Aphis craccivora*	e.	Rice Yellow dwarf

12)

1.	*Diaphorina citri*	a.	Yellow vein mosaic of okra
2.	*Dialeurodes citri*	b.	Citrus dieback
3.	*Tylenchulus semipenetrans*	c.	Kolshi disease of citrus
4.	*Bemisia tabaci*	d.	Tomato spotted wilt virus of groundnut
5.	*Caliothrips indicus*	e.	Greening melody of citrus

13)

1.	*Pentalonia nigronervosa*	a.	Coffee swollen shoot
2.	*Bemisia tabaci*	b.	Papaya leaf curl
3.	*Tetranychus spp.*	c.	Banana bunchy top
4.	*Ferrisia virgata*	d.	Papaya masaic
5.	*Myzus persicae*	e.	Chilli leaf curl

14)

1.	*Sitophilus oryzae*	a.	Tenebrionidae
2.	*Oryzaephilos surinamensis*	b.	Curculionidae
3.	*Tribolium castanium*	c.	Cucujidae
4.	*Sitotroga cereallella*	d.	Gallariidae
5.	*Corcyra cephalonica*	e.	Gelechiidae

15)

1.	*Callosobruchus chinensis*	a.	Bostrychidae
2.	*Rhizopertha dominica*	b.	Bruchidae
3.	*Trogoderma granarium*	c.	Lygaeidae
4.	*Aproaerema modicella*	d.	Gelechidae
5.	*Elasmolemus sordidus*	e.	Dermestidae

16)

1.	*Asphondaylia sesame*	a.	Acrididae
2.	*Helopeltis antonii*	b.	Cecidomyidae
3.	*Antherigona soccata*	c.	Capsidae
4.	*Mythimna separata*	d.	Anthomyidae
5.	*Hieroglyphus banian*	e.	Noctuidae

17)

1.	*Leptocoris acuta*	a.	Aphididae
2.	*Pelopidas mathias*	b.	Cecidomyidae
3.	*Rhophalosiphum maidis*	c.	Coreidae
4.	*Colemenia sphenerioides*	d.	Acrididae
5.	*Contarenia sorghicola*	e.	Hesperiidae

18)

1.	*Amsacta moorei*	a.	Lymantriidae
2.	*Mylabris pustulata*	b.	Pyrrhocoridae
3.	*Bemesia tabaci*	c.	Eriophydae
4.	*Aceria gossypii*	d.	Aleurodidae
5.	*Dysdercus cingulatus*	e.	Meloidae

19)

1.	*Oxycaraenus hyalipennis*	a.	Pterophoridae
2.	*Exelastis atomosa*	b.	Lygaeidae
3.	*Melanagromyza obtuse*	c.	Gelechidae
4.	*Phthorimaea operculella*	d.	Pyralidae
5.	*Leucinodes orbonallis*	e.	Agromyzidae

20)

1.	Leaf minor	a.	Agromyzidae
2.	Lemon butterfly	b.	Scarabidae
3.	Scale insect	c.	Papilionidae
4.	White grub	d.	Coccidae
5.	Locust	e.	Acrididae

21)

1.	*Ophiomyia phaseoli*	a.	Agromyzidae
2.	*Urentius sentis*	b.	Pentatomidae
3.	*Bagrada cruciferae*	c.	Tingidae
4.	*Amrasca biguttulla biguttulla*	d.	Cicadallidae
5.	*Amritodus atkinsoni*	e.	Jassidae

22)

1.	Diamond back moth	a.	Tenthredinidae
2.	Mustard sawfly	b.	Aleurodidae
3.	Black fly	c.	Trypetidae
4.	Ber fruit fly	d.	Noctuidae
5.	Cutworm	e.	Plutellidae

23)

1.	Cabbage butterfly	a.	Pieridae
2.	Yellow pumpkin beetle	b.	Chrysomelidae
3.	Blister beetle	c.	Meloidae
4.	Mealy bug	d.	Pseudococcidae
5.	Red ants	e.	Formicidae

24)

1.	*Orseolia oryzae*	a.	Carposinidae
2.	*Nephotettix virescence*	b.	Cicadellidae
3.	*Aspidiotus destructor*	c.	Coccidae
4.	*Batocera rubus*	d.	Cerambycidae
5.	*Meridarchis scyrodes*	e.	Cecidomyidae

25)

1.	*Carpomyia vesuviana*	a.	Trypetidae
2.	*Deudorix isocrates*	b.	Curculionidae
3.	*Drosicha mangiferae*	c.	Pseudococcidae
4.	*Rhynchophorus ferrugineus*	d.	Lycaenidae
5.	*Oryctes rhinoceros*	e.	Dynastidae

26)

1.	*Nephopteryx eugraphella*	a.	Coccidae
2.	*Agrius convolvuli*	b.	Pyralidae
3.	*Cylas formicarious*	c.	Sphingidae
4.	*Holotrichia serrata*	d.	Apionidae
5.	*Melanaspis glomerate*	e.	Scarabaeidae

27)

1.	*Sesamia inference*	a.	Termitidae
2.	*Pyrilla perpusilla*	b.	Fulgoridae
3.	*Emmalocera depressella*	c.	Aphididae
4.	*Microtermes obsesi*	d.	Noctuidae
5.	*Pentalonia nigronervosa*	e.	Pyrallidae

28)

1.	*Chillo partellus*	a.	Delphacidae
2.	*Nilparavatta lugens*	b.	Pyrallidae
3.	*Epilachna vigintoctopunctata*	c.	Curculionidae
4.	*Sternochetus mangiferae*	d.	Tenthoidinidae
5.	*Bactrocera cucurbitae*	e.	Coccinellidae

29)

1.	*Raphidopalpa faveicollis*	a.	Chrysomelidae
2.	*Clavigralla gibbosa*	b.	Agromyzidae
3.	*Melanagromyza obtusa*	c.	Coreidae
4.	*Sylepta derogata*	d.	Pyralidae
5.	*Myllocerus maculosus*	e.	Curculionidae

30)

1.	*Scirtothrips dorsalis*	a.	Thripidae
2.	*Earias vitella*	b.	Noctuidae
3.	*Pectinophora gossypiella*	c.	Gelechidae
4.	*Opisina arenosella*	d.	Xylorictidae
5.	*Phyllocnistis citrella*	e.	Gracillaridae

31)

1.	*Diaphorina citri*	a.	Muridae
2.	*Eudocima fullonica*	b.	Psylidae
3.	*Dicladispa armigera*	c.	Noctuidae
4.	*Antherigona socatta*	d.	Chrysomelidae
5.	*Rattus rattus*	e.	Muscidae

32)

1.	Onychophora	a.	Centipedes
2.	Chilopoda	b.	Milliped
3.	Diplopoda	c.	Prawn, Crab and Wood louse
4.	Crustacea	d.	Insects
5.	Hexapoda or Insecta	e.	Peripatus

33)

1.	Tomato leaf curl	a.	*Thrips tabaci*
2.	Tomato spotted wilt	b.	*Thrips tabaci*
3.	Papaya leaf curl	c.	White fly
4.	Tobacco leaf curl	d.	White fly
5.	Bhendi yellow vein mosaic	e.	*Bemisia tabaci*

34)

1. Cowpea mosaic	a. *Cestius phycitis*
2. Little leaf of Brinjal	b. *Aphis craccivora*
3. Lleishmaniasis	c. Sand Fly
4. *Myzus persicae*	d. *Thrips tabaci*
5. Tomato spoted wilt	e. Tobbaco rosettese

35)

1. Kala-azar	a. Black flies or *Simulian* spp.
2. *Anopheles mosquito*	b. Malarial fever
3. *Human onchocerciasis*	c. *Phlebotomus* spp. (Sand Fly)
4. Banana bunchy top	d. *Pentalonia nigronervosa*
5. Papaya mosaic	e. *Aphis gossypii*

36)

1. Rice yellow dwarf	a. *Nephotettix virescens*
2. Tungro disease of Rice	b. *N. nigropictes*
3. Bubonic plague	c. Rat flea, *Xenopsylla cheopis*
4. Citrus tristeza	d. *Aphis gossypii*
5. Chilli mosaic	e. Aphid (*Toxoptera citricida*)

35)

1. *Sesamum phyllody*	a. *Orosius albicinctus*
2. Banana Steak Disease	b. Mealy bug
3. Cardamom mosaic (Katte disease)	c. *Pentalonia nigronervosa*
4. Coconut root wilt	d. *Stephanitis typicus*
5. Dengue	e. *Aedes aegypti*

36)

1. Father of Taxonomy	a. Agostino Maria
2. Father of Insect Physiology	b. Wigglesworth
3. Father of Biology	c. Megnin
4. Father of Forensic Entomology	d. Aristotle
5. Father of Insect Pathology	e. Carl Linnaeus's

37)

1.	Father of Host Plant Resistance in entomology	a.	R. H. Painter
2.	1st Govt. Entomologist of Madras state	b.	T. B. Fletcher
3.	1st Emperical Forest Entomologist	c.	E. P. Stebbing
4.	Inventer of Biometer	d.	S. Pradhan
5.	1st Emperical Entomologist	e.	H. M. Lefroy

38)

1.	Indian Institute of Natural Resins and Gum's	a.	Ranchi
2.	Indian Lac Research Institute	b.	1939
3.	Locust Warning Organisation	c.	Ranchi (1925)
4.	Imperical Forest Research Institute	d.	Pune, 1937
5.	All India Bee-Keepers Association	e.	Dehradun, 1916

39)

1.	Commonwealth Institute of Biological Control (CIBC)	a.	Trininad (West Indies)
2.	National Bureau of Agriculturally Important Insect	b.	Bangalore
3.	Entomological Society of India	c.	1938
4.	Central Sericultural Germplasm Resources Centre	d.	Hasur (Tamil Nadu)
5.	National Bureau of Agriculturally important Microorganisum	e.	Mau Nath Bhanjan (U.P.)

40)

1.	Central Sericultural Research and Training Institute	a.	Mysore (Karnataka)
2.	Central Sericulture Research Station for Mulbery Silkworm	b.	Namkum, Ranchi
3.	Central Tasar Research Station	c.	Pune
4.	Central Honey bee Research and Training Institute	d.	Ranchi
5.	Lac Research institute or Indian Institute of Natural Resins and Gum's	e.	Berhampur, West Bengal

Answer Key

1)	1- c	2- b	3- a	4- e	5- d
2)	1- e	2- d	3-c	4-b	5- a
3)	1- c	2- a	3- b	4- e	5- d
4)	1- d	2- e	3- b	4- a	5- c
5)	1- b	2- c	3- d	4- e	5- a
6)	1- e	2- b	3- d	4- c	5- a
7)	1- e	2- e	3- d	4- c	5- b
8)	1- e	2- c	3- d	4- a	5- b
9)	1-a	2- b	3- e	4- d	5- c
10)	1- b	2- a	3- d	4- c	5- e
11)	1- c	2- b	3- a	4- e	5- d
12)	1- e	2- c	3- b	4- a	5- d
13)	1- c	2- b	3- e	4- a	5- d
14)	1- b	2- c	3- a	4- e	5- d
15)	1- b	2- a	3- e	4- d	5- c
16)	1- b	2- c	3- d	4- e	5- a
17)	1- c	2- e	3- a	4- d	5- b
18)	1- a	2- e	3- d	4- c	5- b
19)	1- b	2- a	3- e	4- c	5- d
20)	1- a	2- c	3- d	4- b	5- e
21)	1- a	2- c	3- b	4- e	5- d
22)	1- e	2- a	3- b	4- c	5- d
23)	1- a	2- b	3- c	4- d	5- e
24)	1- e	2- b	3- c	4- d	5- a
25)	1- a	2- d	3- c	4- b	5- e
26)	1- b	2- c	3- d	4- e	5- a
27)	1- d	2- b	3- e	4- a	5- c
28)	1- b	2- a	3- e	4- c	5- d
29)	1- a	2- c	3- b	4- d	5- e
30)	1- a	2- b	3- c	4- d	5- e
31)	1- b	2- c	3- d	4- e	5- a
32)	1- e	2- a	3- b	4- c	5- d

33)	1- a	2- b	3- c	4- d	5- e
34)	1- b	2- a	3- c	4- e	5- d
35)	1- c	2- b	3- a	4- d	5- e
36)	1- b	2- a	3- c	4- e	5- d
37)	1- a	2- b	3- c	4- d	5- e
38)	1- e	2- b	3- d	4- c	5- a
39)	1- a	2- b	3- c	4- d	5- e
40)	1- a	2- c	3- b	4- e	5- d
41)	1- a	2- b	3- c	4- d	5- e
42)	1- a	2- e	3- d	4- c	5- b

List of Various Disease with their Insect Vector

Insect Pests		*Disease*
Common Name	*Scientific Name*	
Green leaf hopper	*Nephotettix virescens*	Rice grassy stunt, rice transitory yellowing, rice tungro, rice yellow orange leaf *
	Nephotettix nigropictus	Rice yellow dwarf
Zig zag leaf hopper	*Recilia dorsalis*	Rice yellow dwarf, rice orange leaf, rice tungro, rice yellow orange leaf
Brown plant hopper	*Nilaparvata lugens*	Rice ragged stunt
Plant hopper	*Tagosodes orizicolus, Sogatodes oryzicola*	Rice hoja blanca
Leaf hopper	*Orosius albisinctus*	Sesame phyllody
Sorghum shoot bug	*Peregrines maidis*	Corn mosaic, sorghum chlorosis, freckled yellows of sorghum
Ragi leaf hopper	*Cicadulina bipunctella bipunctella*	Ragi mosaic
White fly	*Bemisia tabaci*	Bhendi yellow vein mosaic, chilli mosaic, pulses yellow mosaic, cotton mosaic, tapioca mosaic, tomato spotted wilt, cotton leaf curl, tobacco leaf curl
Aphid	*Pentalonia nigronervosa*	Banana bunchy top, banana mosaic, foorkey disease of cardamum, katte disease of cardamum

Insect Pests		*Disease*
Common Name	*Scientific Name*	
	Myzus persicae	Potato leaf roll, potato virus x, cucumber mosaic, sugar beet yellow mosaic
	Aphis gossypii	Cucumber mosaic
	Aphis craccivora	Groundnut rosette virus, cow pea mosaic, papaya mosaic
	Aphis sacchari, Aphis maidis	Augarcane grassy shoot
	Rhopalosiphum maidis	Sugarcane mosaic
	Toxoptera aurantii, T. citricidus	Citrus tristeza virus
	Acrythosiphon dirhodum	Barley yellow dwarf
	Acrythosiphon pisum	Bean yellow mosaic
Thrips	*Scirtothrips dorsalis*	Chilli leaf curl
	Frankliniellan schutzei	Soybean spotted wilt
	Frankliniellan schutzei, Thrips palmi	Groundnut bud necrosis virus
	Thrips tabaci	Tomoto spotted wilt
Mealy bug	*Saccharcoccus sacchari*	Sugarcane mottling
	Dysmicoccus brevipes	Pine apple root wilt
	Planococcoides citri	Cacao mottle leaf
	Planococcoides njalensis	Cacao swollen shoot
Psyllid	*Diaphorina citri*	Citrus greening
Eriophyid mite	*Aceria tulipae*	Wheat streak mosaic, wheat spot mosaic
	Aceria cajani	Redgram sterility mosaic
Mango bud mite	*Aceria mangiferar*	Mango malformation
Elm bark beetle	*Scolytus multistriaus*	Dutch elm disease

* More than one vector for diseases.

Chapter 17
Memory Based Questions

(ICAR-JRF EXAMINATION)
ENTOMOLOGY

1. **Stone fly is the indicator of________.**
   Ans: Air pollution/**water pollution**/both/none

2. **In India first Report of control of white ant ________.**
   Ans: Bacteria/fungi/virus/**red ant**

3. **Green lace wing feeds on________.**
   Ans: Adult/pupa/**caterpillar**/all

4. ***Eublema amabillis* feeds on________.**
   Ans: Mulberry leaf/ber leaf/**lac insect**/caster leaf

5. **Which stage of leaf miner bore in leaf?**
   Ans: Adult/larvae/**maggot**/none

6. **Bombay Natural History Society established in________.**
   Ans: **1883**

7. **Haemolymph percentage in insect is?**
   Ans: **30 to 80**/15to75/3to40/none

8. **Which of the following is Dipterous leaf miner?**
   Ans: **Pea**/groundnut/citrus leaf miner/none

9. **Which of the following aphid fed on pulses?**
   Ans: ***Aphis craccivora***

10. **Sugarcane mosaic is transmitted by________.**
    Ans: ***Rophalosiphum maidis****/M.persicae/A.gossypii*/none

11. **Axillary sclerite absent in________.**
    Ans: **Pdonata**

12. **In which of the following pupal stage present in male but absent in female?**
    Ans: **Strepsiptera**

13. **Glyphosate is a________.**
    Ans: **Herbicide**/insecticide/growth regulator/none

14. **Which of the following is antitranspirant?**
    Ans: **PMA**/kaolin/2, 4-D/none

15. **Wingless ectoparasite on birds and mammals: ________.**
    Ans: **Mallophaga**

16. **In cricket chirping noise produced by ________.**
    Ans: Rubbing of legs together/**wings together**/hind legs and fore wings/fore leg and hind wing

17. **Luminous is the characteristic of the order________.**
    Ans: **Coleoptera**

18. **Which of the following is pest?**
    Ans: Epilachna vigintioctopunctata

19. **Anamorphism is present in________.**
    Ans: **Protura**

20. **Monocondylic single mandible present in________.**
    Ans: **Thysanoptera**

21. **Pectinate antennae present in ________.**
    Ans: **Root borer**

22. ***C. cephalonica* belong to family:________.**
    Ans: **Galleridae**

23. **In Lepidoptera pheromone is secreted by________.**
    Ans: **Glandular epidermal cell**

24. **Antenna is absent in________.**
    Ans: **Protura**

25. **Grasping type of leg present in________.**
    Ans: **Preying mantid**

26. **Function of peritropic membrane:________.**
Ans: **Protective**

27. **Cuticle is absent in ________.**
Ans: **Midgut**

28. **1meso, 1meta six abdominal spiracles : ________.**
Ans: **Holopneustic**

29. **Supra-oesophageal gaglion is called as ________.**
Ans: **Brain**

30. **Which of the following endocrine hormone?**
Ans: **Juvenile hormone**

31. **Malphigian tubule is discovered by ________.**
Ans: **Malphigi**

32. **Which of the following excretory organ?**
Ans: Malphigian tubule/integument/alimentary canal/**All**

33. **NBAII located at ________.**
Ans: **Bangalore**

34. **Central Sugarcane Breeding Institute located at ________.**
Ans: **Coimbatore**

35. **Which of the following pest of mango has quartine importance?**
Ans: **Mango Stone weevil and Mango fruit fly**

36. **Which fruit crop has largest area in India?**
Ans: **Mango**

37. **Sugar translocation in sugarcane is depend on________.**
Ans: **Bo**

38. **Mosaic cell membrane theory is given by________.**
Ans: **Nicolson**

39. **Size of aerosol is________.**
Ans: **0.1 to 40 micron**

40. **Fog is produced by using which nozzle?**
Ans: **Thermal**

41. **O_2 supply increases how many times during insect flight?**
Ans: 40/50/100/**200**

42. **Pesticide/metabolites remains in the environment is called as________.**
    Ans: **Residue**

43. **Reproduction in which only male produce is called as________.**
    Ans: **Arrhenotoky**

44. **Avermectin produced from streptomycin is ________.**
    Ans: **Monocyclic lactone**

45. **Area under Bt cotton during 2007-2008 is ________.**
    Ans: **7.6 M ha**

46. **Thiometozam is belongs to________.**
    Ans: **Neonecotinoide**

47. **Which of the following is oxamyl group insecticide?**
    Ans: **Methomyl**

48. **Host of Tassar: ________**
    Ans: ***Shorea robusta***

49. **Symptoms of mango stem borer damage: ________**
    Ans: **Grub feed on bark and branch wilting**

50. **Yellow vein mosaic virus transmitted of bhendi________.**
    Ans: ***B. tabacci***

51. **Introduced pest of coffee ________.**
    Ans: **Berry borer**

52. ***Aphelinus mali* is parasitoids of__________.**
    Ans: **Apple woolly aphid**

53. **Which of the following is larval parasitoids?**
    Ans: **Apenteles**

54. **Which of the following entonopathogenic protozoa of silkworm?**
    Ans: ***Nosema bomycis***

55. **Which of the following trade in maize for resistant to *H. zea*?**
    Ans: **Lethal silk factor**

56. **Token stimuli theory was given by________.**
    Ans: **Frankel**

57. **Popular hive in India is ________.**
    Ans: **Newtone hive**

58. **Efficiency of Bt is highest at pH**
Ans: **9 to 10**

59. **First time in China Bt endotoxin producing gene transferred in which crop?**
Ans: **Tobacco**

60. **Worldwide pest of Apple is ________.**
Ans: **San jose scale**

61. **Which of the following pest damage crop in field and storage condition?**
Ans: **Pulse beetle**

62. **Bushy appearance in sugarcane is due to ________.**
Ans: **Top shoot borer**

63. **Leaf curl of cotton is due to________.**
Ans: **White fly**

64. **Which is first photostable pyrethroid?**
Ans: **Permethrin**

65. **Sting in worker honey bees is modified part of** ovipositor.

66. **Pollination in Mango done by** House fly.

67. **What is mean by Adroconia?**
Ans: **Scent gland**

68. ________**ganglion is known as brain.**
Ans: **Supra-oesophagial**

69. *Nezara viridula* **have________of mouth parts.** Piercing and sucking.

70. **Pair of insecticides banned in India now__________.**
Ans: **DDT + Endosulfan**

71. **Brushing means -** Transferring of larvae on feeding bed.

72. **Rice case worm pupate in** Tubular cases of leaves.

73. **Scientific name of cattle fly** *Hippobosca maculate*.

74. **Honey contains** Vitamin B_{12}.

75. **Glands produces brood feed in worker bees** Pharyngeal gland.

76. **Most toxic chlorinated hydrocarbon** DDT.

77. **Halters are present in** True flies.

78. **Malphighean tubules are absent in** Collombola.

79. **In India bulk of honey derived from** *Aphis dorsata*.

80. **Resistance against all known biotypes is called** Non specific resistance.
81. **In Indian condition which is destructive locust** *Schistocera gregaria.*
82. **Single caterpillar produces silk thread of** 650-1300m.
83. **In which stage fruit fly takes rest** Adult stage.
84. **San jose scale is pest of** Temperate fruits.
85. **Thorax in insects is** Tri segmented.
86. **Bubanic plague disease carried by** Rat Flea.
87. **Which state produce more lac?**
    Ans: **Chattisgarh.**
88. **The light producing substance in Glow worm or fire fly is** Luciferin.
89. **Recent serious and wide spread sucking pest of cotton is** Cotton aphid.
90. **Pest introduced in our country before enforcement of quarantine measures is** *Icerya purchase.*
91. **First abdominal segment fused with metathorax in hymenoptera, this segment is known as** Propodium.
92. **Spiracles absent in** Chironomid larvae.
93. **Order containing predacious insect is** Neuroptera.
94. **Aphid feeding on pulses** *Aphis craccivora.*
95. **Pest of Apple is** *Eriosoma lanigerum.*
96. **___________is parasitic protozoan attacking silkworm** *Nosema bombycis.*
97. **Type of parthenogenesis in which only males are produced** Arrhenotoky.
98. **'Mudgo' variety of paddy resistant to BPH due to** Nutritional deficiency.
99. **Recommended insecticide for controlling termite** Chlorpyriphos.
100. **Most pheromone traps attract _________moths.** Male.
101. **Eri silkworm reared on** Castor.
102. **Laboratory host for multiplication of egg parasitoids is** *Corcyra cephalonica.*
103. **Females are wingless and males with one pair of wings found in** Coccidae.
104. **Immature stages of Ephemeroptera called as** Naiads.
105. **Tibia of hind legs of honey bees modified into** Pollen basket.
106. **Arolium is found between** Claws.
107. **Dorsal region of a thoracic segment known as** Notum.
108. **Molar region of mandibles possess** Blunt teeth.

109. Member of phylum arthropoda have dorsal blood vessel.

110. Insecticide belonging to organophosphatic group malathion.

111. Chronic Anticoagulant used for rat control Warfarin.

112. ________is mammalian pest.

Ans: ***Rattus ratus***

113. Order of head loss is Siphanoptera.

114. Name of insect hormone is ecdysone.

Match the Pairs

1.

(a)	Plutellidae -	*P. xylostella*
(b)	Noctidae -	*H. armigera*
(c)	Pyraustidae -	*L. arbonalis*
(d)	Coccinellidae -	*Epilachna*
(e)	Chrysomelidae -	*Phytotreta cruciferae*

2.

(a)	*L. heteropoda* -	Neristoxin
(b)	*N. tabacum* -	Nicotine
(c)	*C. cinerariaefolium* -	Pyrethum
(d)	*Derris eplicta* -	Rotenone
(e)	Phenolic carbamate -	Carbaryl

3.

(a)	Indian meal moth -	*P. interpunctata*
(b)	Aguimois moth -	*S. cerellela*
(c)	Rice moth -	*C. cephala*
(d)	Almond moth -	*E.cautella*
(e)	Red rust flour beetle -	*T. confusum*

4.

(a)	Aristate -	Housefly
(b)	Pectinate -	Moth
(c)	Silkworm -	Bipectinate
(d)	Monoliform -	Thirps
(e)	Serreted -	Pulse beetle

5.

(a) Grooming - Honey bee

(b) Grasping - Mantide

(c) Jumping - Flea beetle

(d) Digging - Glyllotalpa

(e) Rasping and sucking - Thrips

6.

(a) Rice - *T. incertulus*

(b) Sorghum - *A. soccata*

(c) Maize - *T. indicus*

(d) Pigeon Pea - *E. atomosa*

(e) Mustard - *A. proxima*

(JRF ENTOMOLOGY and NEMATOLOGY 2010-2011)

1. **Which nematode transfered by insect?**
   Ans: ***Rhadinaphelenchus cocophilus.***

2. **Three segmented stylet with median bulb present in________.**
   Ans: **Tylenchida.**

3. **Stomatostylet present in________.**
   Ans: **Tylenchida/**Dorylaimada/aphelenchida/None.

4. **Which of the following is pest of mushroom?**
   *Ans: D.dipsaci/D.destructor/D.angustatus/**D. myceliophagous.***

5. **Fan leaf virus trasfered by ________.**
   Ans: **Xiphinema/**Trichodorus/Longidorus/None.

6. **Which of the following mushroom nematode?**
   *Ans: **D. myceliophagous**/D.disaci/D.destructor/All.*

7. **Infective stage of root knot nematode________.**
   Ans: **J2/**J3/J4/Adult.

8. **Infective stage of reiniform nematode________.**
   Ans: juvenile/**pre-adult**/mature/all

9. **ETL of root knot nematode (juvenile/gm soil) ________.**
   Ans: 4/3/2/5

10. **Rajkiran variety of barley resistant to ________.**
Ans: ***H.avenae*** */ RKN/*Both / None

11. **Race-B of reiniform nematode infect which crop**
Ans: cowpea / cotton / caster / All

12. **Which of the following nematode infest G.nut**
*Ans: M.halpa/M.icognita/M.exigua/**M.arenaria***

13. **Which root knot nematode does not form gall**
*Ans: M. halpa/M. icognita/**M. nassi**/M. tricycli*

14. **Which of the following entomopathogenic nematode**
Ans: **steinernema/**xenorhabditus / photorhabditus / All

15. **White tip nematode is________.**
*Ans: A. fragariae/A. ritzemabosi/**A. bessi*** / All

16. **Vascular wilt of banana is caused by nematode with**
Ans: ***Fusarium oxysporum f. sp. Cubense****/Rhizotonia solani/*Both (a) and (b) / none

17. **Bacteria associated with entomopathogenic nematode**
Ans: xenorhabditus / photorhabditus / **Both (a) and (b)/**steinernema

18. **In potato cyst nematode which fenestrae is**
Ans: uni / bi / amphi / **none (circum)**

19. **Perineal pattern is used for identification of which nematode**
Ans: **Meloidogyne** / Heterodera / Globodera / All

20. **One gonoid present at posterior side is called**
Ans: prodelphic / didelphic / **opisthodelphic/**none

20. **In nematode which of following system absent**
Ans: reproductive / excretory / nervous / **respiratory**

21. **Cowpea variety resistance to *Meloidogyne***
Ans: ML-62 / ML-80 / Nemamukta / C-142

22. **Pine wilt nematode is ________**
Ans: Xiphinema / ***Bursaphelenchus xylocophilus*** Longidorus / None

23. **Rice root nematode is**
*Ans: **Hirshmaniella oryzae**/pratylenlus/rotylenchulus/None*

24. **False root knot nematode is**
Ans: ***Noccobus*** **spp.**

25. Virus tranmitting nematode belong to which order
Ans: Monochida / **Dorylaimada** / Spicurina / Tylenchida

26. Anhydrobiotic stage of *Ditylenchus* is called as
Ans: Eelworm / **eelwool** / nemacyst / none

27. NETU virus transmitted by
Ans: Xiphinema/**Trichodorus** / Longidorus / paralongidorus

28. Sexual diamorphism present in
Ans: **Meloidogyne**

29. Rectal gland present in which nematode?
Ans: **Meloidogyne**

30. Glandular excretory system present in
Ans: **Adenophorea**

31. Nurse cell produced by
Ans: ***Tylenchulus semipenetrance***

32. caudal alae present in
Ans: **Meloidogyne**

33. How many times nematode moult
Ans: **4**

34. Which of coelom present in nematode
Ans: **Pseudocoelom**

Match the Pairs

1.

(a) Dagger -	Xiphinema
(b) Sting -	Belonolaimus
(c) Kala Hista -	Aphelenchoides
(d) Mushroom -	*D. myceliphagous*
(e) Ear cockle -	*A. tritici*

2.

(a) Pralinage -	Application of nematicide with clay slurry
(b) Solarization -	Cultural method
(c) Trizanine -	Nemacure
(d) Human nematode -	Ascaris
(e) Phytomicide -	Chemical secreted by higher plants

3.

(a)	Paratylenchus -	Pin nematode
(b)	Pratylenchus -	Lesion nematode
(c)	Hoplolaimus -	Lance nematode
(d)	Longidorus -	Neeedle nematode
(e)	Tylenchorhynchus -	Stunt nematode

www.ingramcontent.com/pod-product-compliance
Lightning Source LLC
Chambersburg PA
CBHW070252260926
51849CB00001BA/236
9789351302261